AF771306

Fungicides in Crop Protection

Fungicides in Crop Protection

Editor

Harshad Khuspe

scitus
academics

Fungicides in Crop Protection

Edited by **Harshad Khuspe**

Printed in 2017

ISBN: 978-1-68117-066-4

Library of Congress Control Number: 2015935464

© 2016 by
SCITUS Academics LLC,
616, Corporate Way, Suite 2, 4766,
Valley Cottage, NY 10989

www.scitusacademics.com

This book contains information obtained from highly regarded resources. Copyright for individual articles remains with the authors as indicated. All chapters are distributed under the terms of the Creative Commons Attribution License, which permits unrestricted use, distribution, and reproduction in any medium, provided the original author and source are credited.

Notice

Reasonable efforts have been made to publish reliable data and views articulated in the chapters are those of the individual contributors, and not necessarily those of the editors or publishers. Editors or publishers are not responsible for the accuracy of the information in the published chapters or consequences of their use. The publisher believes no responsibility for any damage or grievance to the persons or property arising out of the use of any materials, instructions, methods or thoughts in the book. The editors and the publisher have attempted to trace the copyright holders of all material reproduced in this publication and apologize to copyright holders if permission has not been obtained. If any copyright holder has not been acknowledged, please write to us so we may rectify.

Contents

Preface

Plant pathogenic fungi cause devastating damage to crop production worldwide. The growing global population necessitates reduced crop losses to improve food security, and the control of fungal plant pathogens is vital to help maintain food production. Providing a concise and balanced review of fungicides used in crop protection, this book describes the science of fungicide use, selection and resistance within the context of farming situations. Major updates and additions reflecting the emergence of two new classes of fungicides and the increased incidence of fungicide resistance are included in this new edition, which also discusses legislative requirements to reduce fungicide applications, and current trends in fungicide use. Plant protection products play a crucial role in maintaining adequate supplies of high quality, affordable food. Without these products to keep weed, pest and disease pressures in check, crop yields would fall by around a third. Plant protection products are also widely used outside crop production, for example in maintaining the quality of gardens, golf courses and sports pitches, and ensuring the safety of our railways and roads. The plant protection products are lifesaving medicines for plants. When attacked by disease or pests, a plant's natural response is to produce its own chemical protection. Man-made pesticides are only used when a plant's own chemical defenses do not work well enough. But like human and veterinary medicines, plant protection products can pose some risks if they are not used with due care. Their registration and use is therefore strictly regulated and controlled. Plant protection products help farmers protect their crops against yield losses and damage caused by weeds, diseases and insects. Keeping problem pests under control is essential to provide a sustainable supply of safe, affordable, wholesome food.

Editor

Chapter 1

Detection of Fungicide Resistance

Janna L. Beckerman

Department of Botany and Plant Pathology, Purdue University, West Lafayette, USA

INTRODUCTION

Fungicide resistance is a form of selection that describes a fungus's ability to survive and reproduce in the presence of a fungicide[1] - . Practical resistance (field resistance) results when the prevalence of fungicide resistant isolates reaches a critical threshold where disease control is no longer observed. The primary factors that select for fungicide resistance in an organism are:

1 the biology of the pathogen,

2 the mechanism(s) of action of the fungicide,

3 the rate and frequency of fungicide application.

[1] - Due to issues of length, brevity and complexity, this review will only deal with fungicides that are for the control of true fungi, and not fungicides for the management of Chromista.

It is important to note that most fungi show a broad range of sensitivities to the spectrum of different fungicides. For example, most fungicides for the control of Chromista (e.g., mefenoxam, fluopicolide, dimethomorph) do not control true fungi; some fungicides are specific for ascomycete (e.g., thiophanate-methyl) or basidiomycete fungi (flutolanil), and species within the same genus of fungi may respond differently to certain fungicides (Leroux et al. 2012). Additionally, different genotypes exist within a spectrum of sensitive to resistant (Albertini et al. 1999). The relationship between the frequency of fungicide application and resistance has been established (King and Griffin, 1985; Suzuki et al. 2010), and the role of rate or dose in this process is deserving of further study (Genet et al. 2006;van den Bosche et al. 2011).

In 2006, the world market consumed approximately 520 million pounds of fungicide (Grube et al. 2011), a number that is expected to increase as the consumption of other pesticides declines (Troy, 2011). The discovery, development, and registration of a new pesticide comes at a cost (not including the capital costs or production). In 2006, this price was estimated to exceed $180 million (Whitford et al. 2006). With fewer fungicides available (particularly with new modes of action) and increasing consumption, risk of fungicide resistance is even greater, and failures will more profoundly impact cropping systems. All of these factors increase the need for rapid detection of fungicide resistance. Historically, the detection of fungicide resistance in the field has been difficult because disease control failures can be caused by factors other than resistance, including improper fungicide selection, improper timing, reduction of recommended rate, and erroneous sprayer calibration (Latin, 2011).

Rarely, are molecular proofs of fungicide resistance ever fulfilled with plant pathogenic fungi, although many of the genes involved in fungicide resistance are studied in model systems to elucidate mechanisms (Zhang et al. 2002). Usually, resistant individuals are detected using bioassays (sensitivity tests or poison plate assays) and those resistant isolates may be correlated with molecular markers, many of which are based upon previous studies (Koernraadt et al. 1992; Albertini et al. 1999; Schnabel and Jones, 1991; Lesniak et al. 2011). In some instances, segregation analysis between crosses of sensitive and resistance phenotypes is often used to strengthen this correlation (Faretra and Pollastro 1993; Orth et al. 1995; Dyer et al., 2000); however, the ultimate mechanism of resistance is often lacking. Ultimately, the success of molecular detection is only useful if the expected genetic profile that is being screened correlates to the phenotype of resistance. Laboratory tests are required to determine the level of resistance in numerous suspect isolates of the pathogen before an assessment of the status of the orchard, plantation, vineyard or field can be determined. Traditionally, this meant direct-plating of single-spore isolates or mycelial plugs on medium amended with various concentrations of fungicides under specified growth conditions, and then determining inhibition of growth and/or spore germination (Koller et al. 1997). Many fungi, including the economically important downy mildews, powdery mildews, and rust, cannot be cultured. Other fungi, like *Venturia inaequalis,* grow slowly in culture, taking at least four weeks to obtain results, which, when achieved, are no longer useful for in-season disease management recommendations (Chapman et al. 2011, and others). Unfortunately, the phenotypic comparison of fungicide resistance can be inconsistent due to media choice (Cox et al. 2009; Rampersad 2011), choice of active ingredient (technical grade versus field fungicide), degradation of material, stability of resistance (Zhu et al. 2012; Cox et al. 2007), and genetic background of isolates. One method of circumventing this inconsistency is through the direct testing of genotype, which has resulted in an emerging paradigm of nucleic acid-based detection systems for the rapid identification of fungicide resistance.

Molecular methods have the potential to provide a more rapid and reliable assessment of fungicide resistance, and there are many examples of the successful use of these methods in applied plant pathology (for a review, see Vincelli and Tisserat, 2008). Genetic testing of potentially resistant isolates can be performed directly from disease lesions, obviating the need for isolation, and subsequent growth of the fungus *in vitro* (Quello et al. 2009; others). Furthermore, molecular methods characterize genotype, not phenotypic expression under laboratory conditions. This is particularly important in the case of recalcitrant fungal pathogens, like *V. inaequalis*or *Mycosphaerella fijiensis* and obligate pathogens such as *Blumeria* and *Plasmopara*. In the case of obligate pathogens in particular, genotype can be readily identified long before phenotype can be determined due to the slow growth of the organism.

As a paradigm, nucleic acid-based detection systems offer rapid (within hours) and sensitive (to picograms) methods to detect the presence of alleles known to confer resistance. First used byKoenraadt et al. (1992b) to detect benomyl resistance in *V. inaequalis,* a variety of sensitive and sophisticated nucleic acid-based detection systems have since been developed and promoted to identify resistance.

However, it is impossible to review the developing paradigm and not question how we implement these detection systems to provide the end-user, in this instance, the farmer, with the information necessary to make the appropriate management decision. The objective of this review is to briefly discuss the current fundamental approaches of nucleic acid based systems, and the currently known targets for the molecular detection of fungicide resistance. The paper concludes with the limitation of these techniques, the impact detection has had on the management of fungicide resistance, and future directions.

NUCLEIC ACID BASED DETECTION TECHNIQUES

The molecular detection of fungicide resistance can be boiled down into three fundamental techniques: Hybridization, amplification, and sequencing. It is important to stress though, that most amplification technologies used today are also partly based on hybridization technology, and that all sequencing technology is based upon amplification. This article is not designed to review the preponderance of available techniques; for that information, the reader is encouraged to review the current literature, as new application technologies are produced every year. Instead, this article focuses on certain paradigms that have developed between laboratories for the detection of fungicide resistance, and to provide both structure and context as to where we are currently positioned.

HYBRIDIZATION

Hybridization is one of the oldest molecular techniques: Sample DNA is denatured into single strands and allowed to anneal with a single-stranded probe labeled with some type of signal (radioactive isotopes, antibodies, enzymes or chemiluminescent compounds) to permit detection. Target DNA is bound to solid support (historically nitrocellulose then nylon, although today numerous substrates are available, including magnetic beads and polystyrene microspheres). Direct hybridization is the simplest assay for single nucleotide discrimination. For 15 to 20-base oligonucleotides, the approximate melting temperature for hybridization of a perfectly matched template compared to one with a single base mismatch can differ by several degrees (Ikuta et al. 1987)—a fact that can be exploited to create a variety of multiplex detection techniques, and discriminate within samples or between samples to a single nucleotide polymorphism.

AMPLIFICATION

Briefly, polymerase chain reaction (PCR) is the basis for all amplification-type reactions and involves the heating of the sample DNA for denaturing, followed by the annealing of the small, oligonucleotides that serve as primers for DNA polymerase, followed by the extension of the primers by a thermostable, DNA polymerase. Numerous books and laboratory manuals exist as references for PCR operation and optimization (Sambrook et al, 1989). PCR based detection of fungicide resistance depends upon the ability of the reaction to selectively amplify specific regions of DNA, and usually require several post-PCR steps, including agarose gel electrophoresis for either confirmation of amplicon presence or size, or restriction enzyme analysis (Lesniak et al. 2011; Fontaine et al. 2009;Quello et al. 2009; and reviewed by Ma and Michailides 2005). The development of new fluorescent techniques (LAMP, etc) has led to novel assay formats that greatly simplify the protocols used for the detection of specific nucleic acid sequences (Nurmi et al 2000) and allow for the detection of a specific PCR product in a homogeneous solution without the need to open the amplification tubes after PCR or gel electrophoresis (Tomlinson et al. 2012). In these techniques and their variations, PCR products are monitored as they are generated during the course of the reaction via one of two ways: By fluorescent or chemiluminescent dyes that bind to double-stranded DNA in a nonspecific fashion, or by fluorescence-labeled probes that bind to specific sequences. As a result, PCR amplification, amplicon detection and analysis are all achieved in a single reaction (Figure 1). If this is not enough detection power, multiple, sequence-specific probes with unique fluorescent reporters can be added to the reaction, allowing for additional, and simultaneous, determination of multiple products.

Techniques such as this are ideally suited for the detection of fungicide resistance, particularly if multiple alleles are involved. Furthermore, the results can be read in real time as the PCR product accumulates or at the end of the thermal cycling protocol directly from the amplification wells. Although many scientists believe that the choice between real time or "standard" PCR (and gel electrophoresis) depends on whether a quantitative or qualitative assay is desired (Nurmi et al. 2000), the reality is that equipment expense and laboratory expertise limits most "applied" labs, resulting in a preponderance of scientifically dazzling techniques that may be used in a human clinical setting, or a plant pathology laboratory focused on basic science, but are rarely, if ever, subsequently tested using isolates from a field failure.

HYBRID TECHNOLOGIES

Technologies such as Luminex Xmap powerfully combine hybridization and PCR to create a technique that is capable of discriminating and reporting up to 500 different reactions in a single reaction vessel in just a few seconds per sample (Dunbar 2005; Luminex 2011). The approximate melting temperature for 15 to 20-base oligonucleotides can differ by several degrees compared to a "perfectly matched" hybridization (Ikuta et al. 1987). By exploiting and combining this discriminatory hybridization temperature with PCR and microarrays systems and digital imaging, the user can screen up to several hundred thousands of DNA probes (either PCR products or synthetic oligonucleotides) per square centimeter of a solid matrix. Multiple readings per beadset provide built-in internal controls (Luminex 2011). For medically relevant human pathogens, whole genome arrays have been developed; it is hardly a stretch to imagine development of arrays for a few dozen alleles for important genes, or even genes important for fungicide resistance. Although financially out of reach for most labs, few could envision the sequencing of entire genomes for a few thousand dollars as we do today, and it is hopefully only a matter of time before microarrays to detect fungicide resistance in agriculturally important crops are widely available. Today, and certainly for the next few years, both cost and accessibility also remain obstacles to the development of such arrays. Within reason and immediate reach is a 96-well format that provides fast, simple, and highly reproducible analyses of up to 96

PCR products—which still translates into an assortment of alleles for a variety of fungicide resistance genes.

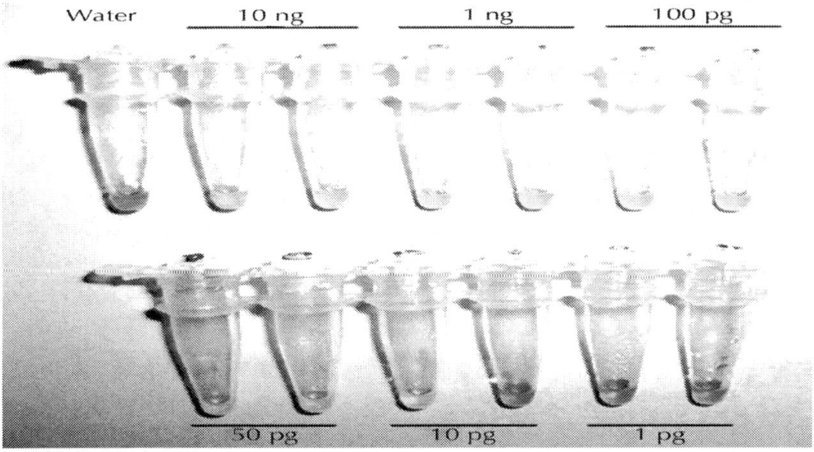

Figure 1: LAMP technologies have been used to detect the presence of invasive species, like *Phytophthora ramorum*. The presence of LAMP product in positive reaction mixtures causes a color change from orange to yellow. In time, this technology could be used to detect some type of fungicide resistance in the field. Photo from Tomlinson et al. 2007.

For most, smaller laboratories, a more realistic approach concerns the use of polymerase chain reaction coupled with cleaved amplified polymorphic sequences (PCR-CAPS). Restriction-fragment length polymorphisms due to small nucleotide polymorphisms (SNP) that co-segregate or are caused by fungicide resistance create or abolish restriction sites in PCR products, and can be exploited for detection of fungicide resistance through the careful selection of locus-specific oligonucleotide primers (Banno et al. 2008; Lesemann et al. 2007, Quello et al. 2009, Fontaine et al. 2009; others).

Assuming that a SNP is present and can be used, oligonucleotide primers with unique sequences are used to amplify a defined locus, followed by the use of a restriction enzyme that can discriminate between resistant and sensitive isolates (Figure 2). An alternative to PCR-CAPS is allele specific (AS-PCR), in which the mutation that confers resistance is used to design primers that specifically amplify the mutated allele, but not the wild-type one (Fontaine et al. 2009,

Lesniak et al. 2011, others). This approach is more sensitive, and does not require a RFLP to detect resistance, but also does not detect any heteroplasmy or heterozygosity.

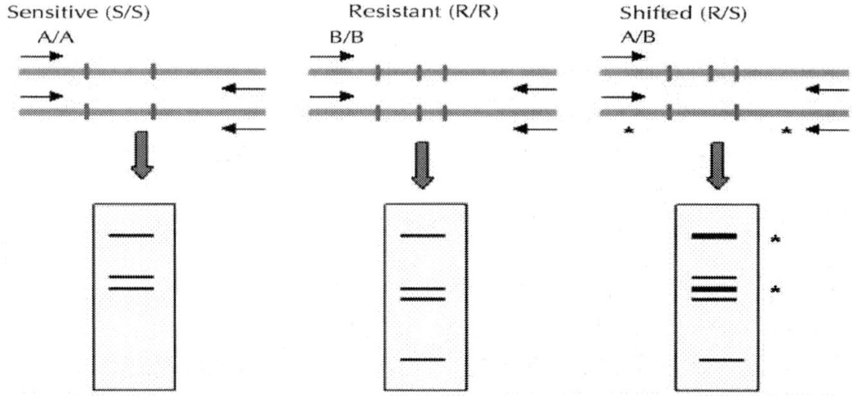

Figure 2: Cleaved Amplified Polymorphic Sequences (CAPS) polymorphisms result from single nucleotide polymorphisms (SNPS), insertions or deletions (INDELS) that create or destroy restriction enzyme recognition sites in polymerase chain reaction (PCR) amplicons. For example, three isolates have different level of fungicide resistance to azoxystrobin: Sensitive S and Resistant R, and shifted (moderately resistant) from the isolate with mitochondrial heteroplasmy R/S. The amplified fragments from S and R contain two and three RE recognition sites, respectively, which is sufficient discrimination in the case of fungi, which are mostly haploid. (Sierotzki et al. 2000; Avenot and Michailides, 2010). This can also result in the identification of heteroplasmy in resistance (as seen inFontaine et al., 2009) with restriction patterns that resemble those found in heterozygous diploids, or even incomplete digestions. When fractionated by agarose or acrylamide gel electrophoresis, the PCR products digested by the RE will give readily distinguishable patterns. Image from http://www.ncbi.nlm.nih.gov/projects/genome/probe/doc/TechCAPS.shtml

DNA SEQUENCING

DNA sequencing, as performed for the last 30 years, has been done via the Sanger method and is the most commonly used sequencing technique available. Genome-sequencing efforts have resulted in technological advances in DNA sequencing and led to the improvement of longer sequencing length, done faster and with less expense. It has also permitted the rapid sequencing of isolates (in part or into total) that did not conform to expected genotype: phenotype relationships (Lesniak et al. 2011; Quello et al. 2010;Leroux et al. 1999, and many others).

Metzker (2009) provides an excellent review of 'next generation sequencing' technology, although the rapid pace of technology marches on. One of the newer sequencing technologies that have immediate application to the detection of fungicide resistance is pyrosequencing. DNA pyrosequencing, a method of sequencing by synthesis, was first introduced in 1996, and is faster and less expensive than traditional Sanger (dideoxy sequencing) DNA sequencing methodologies. Unlike the Sanger method, DNA pyrosequencing utilize a cascade of enzymatic reactions that yield detectable light proportional to incorporated nucleotides. As a result, pyrosequencing yields relatively short read lengths and limited amounts of sequence data per pathogen or microbe. However, with careful target selection and primer placement, DNA pyrosequencing has been used for genotyping, SNP detection, and identification of microbes (Petrosino et al. 2009). Most importantly, pyrosequencing has been used to detect point mutations in antimicrobial resistance genes as a means of molecular resistance testing, including antifungal resistance in clinically important fungi (Wiederhold et al. 2008).

Pyrosequencing offers comparable accuracy to conventional DNA sequencing via the Sanger method, but provides greater opportunity for large sample numbers to be processed in parallel. The reaction is performed in real-time, obviating the need for electrophoresis, labeled nucleotides and primers. This is a technique that can be multiplexed, which economically enables rapid and accurate screening of a large number of samples, however, the prohibitive cost of equipment [$200,000-$1million per machine (Metzker 2009)], and technical expertise limits the use of this technology to research universities and industry.

Finally, a review of the variety of molecular techniques used for diagnostic applications (and the constant development of 'new' techniques) demonstrates that no universal technique exists which is optimal for detection of nucleic acids. The choice of a particular technique is also dependent on the information required, the targets under consideration, and obviously, cost. Regardless of any given or "popular" technique, new techniques continue to be developed which involve new approaches to amplification, hybridization, formats, imaging, and labels.

DETECTION OF FUNGICIDE RESISTANCE

Regardless of the techniques used, the primary targets of fungicide activity (and thus fungicide resistance) have remained remarkably constant, and include mitosis and cell division; sterol biosynthesis; respiration; nucleic acid synthesis; and signal transduction. The widespread use of fungicides has also resulted in resistance in non-target genes, namely drug transporters, which are included in this review due to the role they play in fungicide resistance. The mechanisms of fungicide resistance that have been identified to date, include: i) a mutation in the target site of the antimicrobial agent that reduces the binding of the fungicide; ii) overproduction of the gene that is the target of the fungicide; (iii) reduced uptake of the antimicrobial agent, and (iv) active efflux of the fungicide. Other mechanisms that may have roles in fungicide resistance (based upon host-pathogen interactions, or basic fungal and other microbe biology) include, the possible presence of an enzyme that inactivates the antimicrobial agent (e.g., pisatin-pisatin demethylase) or even mutations that result in the posttranscriptional or posttranslational modification of the target enzyme or other regulatory factor (Mann and Jenson, 2003), resulting in reduced binding of the antimicrobial. Resistance may also be caused by unrecognized mechanisms—a problem for scientists trying to understand mefenoxam, phosphorous acid, or dodine resistance, to name a few of the fungicides where resistance is known, but the primary mechanism of resistance has not yet been identified.

Mitosis and Cell Division: Beta-Tubulin Assembly Inhibitors

Introduced in the 1960s, benzimidazoles were the first penetrant fungicides. Compared to their predecessors, the carbamates, dithiocarbarmates and pthalimides, they were a revolutionary change in fungicides, in that they were noncontact, and extremely effective at low rates. First introduced as benomyl (other MBC fungicides include thiabendazole, and thiophanate-methyl, to name a few), once inside the plant it is metabolized to form methyl benidimidazole carbamate (MBC), which inhibits fungal mitosis via binding to tubulin, the subunit of microtubules essential to forming the mitotic spindle (Ma and Michailides, 2005).

As one of the first noncontact fungicides, it is not surprising that it was one of the first instances of resistance reported, to powdery mildew in greenhouse cucumbers (Schroeder et al. 1969). Since that time, resistance to the benzimidazole class of fungicides has been detected in many fungal species (Ma and Michelides, 2005). Resistance is correlated with point mutations in the β -tubulin gene, with different mutations resulting in altered amino acid changes at the benzimidazole-binding site. These various mutations at different codon sites also result in different levels of resistance: In *V. inaequalis*, mutations at codon 198 resulted in medium resistance, at codon 200 it resulted in very high resistance (Koenraadt et al. 1992a), and at 240 it resulted in low resistance (Quello et al. 2010). In *Monilinia*, only low and high resistance to benomyl and thiophanate-methyl has been observed for field isolates of*M. fructicola* (Ma and Michelides, 2005). Sequence analysis of the β-tubulin gene showed that a single base pair mutation at codon 6 was responsible for the low resistance level to benzimidazoles in all the LR isolates of *M. fructicola* examined. Curiously, different substitutions at the same codon resulted in different degrees of resistance in the cereal eyespot fungus, *Tapesia yallundae* (Pseudocercosporella herpotrichoides): codon changes from Glu to Ala, Gly, Lys, and Gln at position 198 had 50% effective concentration (EC50) values to carbendazim ranging from 0.5 to more than 25 m g/ml in some isolates (Albertini et al., 1999), one of the first instances that demonstrates that differences in genetic background may play a role in expression of specific fungicide resistance genes.

PCR-CAPS was used to determine thiophanate-methyl resistance *Helminthosporium solani* (Cunha and Rizzo, 2005). In *V. inaequalis*, a screen using this approach resulted in the identification of previously unidentified alleles conferring resistance in this fungus, and would have under-reported MBC resistance in 31% of the isolates (Quello et al. 2009) if PCR alone was used for detection. Although the authors continue that a PCR-RFLP based assay may be the best option to screen for fungicide resistance from late-season scab lesions because *V. inaequalis* cannot be reliably cultured due to the application of protectant fungicides for other diseases and competing leaf microflora later in the season, it still leaves open the possibility of new alleles for resistance not being identified, and resistance being under-reported. At this point in time, most apple growers have abandoned thiophanate-methyl for the control of apple scab in the field, although it continues to be used for other diseases. However, these studies are a cautionary tale regarding the sole use of a PCR-based detection to identify fungicide resistance, particularly when no in vitro screening is performed in parallel. This is particularly important as Kawchuk et al. (2002) found that mechanisms other than point mutations in the β -tubulin gene play a role in resistance on *Gibberella pulicaris*; work in the human pathogen,*Candida albicans*, identified the role of a multi-drug resistant transporter (MDR) as responsible for benomyl resistance (Ben-Yaacov et al. 1994). Later work by Sanglard et al. (1995), identified the*BEN*^{R} gene that confers resistance to β -tubulin as an ATP-binding cassette (ABC) transporter (Sanglard et al. 1999).

Detection of specific types of benzimidazole resistance also identifies any negative cross-resistance with diethofencarb (Leroux et al. 1999). The phenylcarbamate diethofencarb was introduced in 1984, and has a similar mode of action to the MBC class of fungicides. In fact, single base pair mutations in codons 198 and 200 result in a readily detectable, negative cross-resistance to diethofencarb (Faretra and Pollastro, 1991; Yarden and Katan, 1993). Negative cross-resistance (NCR) occurs when a novel allele that confers resistance to one toxic chemical results in hyper-sensitivity to another. This mutation therefore results in efficacy to benzimidazole-resistant isolates that possess that allele, but not wild-type isolates. In *B. cinerea*, high levels to resistance carbendazim and thiabendazole, conferred hypersensitivity to diethofencarb, even more so than the benzimidazole sensitive type (Leroux et al. 1999). Furthermore, this negative cross-resistance involved other N-phenylcarbamates and other herbicides that target microtubule assembly (Leroux and Gredt, 1989). Negative cross-resistance was also observed between benzimidazoles and several aromatic hydrocarbon fungicides (e.g., dicloran, OPP (o-phenylphenol)); the phenomenon was first described in cereal eyespot fungus, *T. yallundae,* as well (Leroux and Gredt, 1989), but is more likely due to multi-drug resistance (MDR) mutations than a single structural gene.

SIGNAL TRANSDUCTION: DICARBOXIMIDES AND PHENYLPYRROLES

The dicarboximides are composed of three major products: Iprodione, vinclozolin, and procymidone. The introduction of this class of fungicides coincided with the failure of the benzimidazoles in control of Botrytis in grape. As with benzimidizoles, resistance developed rapidly, due to a combination of concurrent applications of this class without rotation or tank-mixing with other products, coupled with a limited understanding of the process of fungicide resistance. This class of fungicides was primarily used for the control *Botrytis, Alternaria, Sclerotinia* diseases, although they are effective on other pathogens like *Rhizoctonia* spp., and *Fusarium* spp., as well. Not surprisingly, field resistance was first observed in *B. cinerea* (Pommer and Lorenz, 1982), *M. fructicola* (Ritchie, 1983), and *Sclerotinia* spp. (Detweiler et al., 1983).

The molecular mechanisms of dicarboximide resistance involve two separate signal transduction pathways: The two-component histidine kinase and mitogen-activated protein (MAP) kinase cascades. Both MAP kinase and two-component histidine kinase are involved with regulating a diversity of cellular responses including differentiation, cell division, gene expression, heat shock and osmotic response. The first molecular mechanism of dicarboximide resistance was identified in *U. maydis* (Orth et al. 1994), and is one of the few instances where the gene was found to be sufficient for conferring resistance to a wild-type isolate via transformation (Orth et al. 1995). The gene, termed *adr1,* is a 1,218 bp open reading frame with homology to serine/threonine protein kinases that was later identified as the major cAMP dependent protein kinase;

However, later studies by Ramesh et al. (2001), found that the *adr1*-encoded enzyme was not the direct target of vinclozolin inhibition but that mutants with a defect in the regulatory subunit of cAMP-dependent protein kinase *(ubc1)* exhibited resistance to vinclozolin and the aromatic hydrocarbon, chloroneb. Mutants with a defect in the *ubc1*gene also display interesting changes in morphology, including a reduction in multiple budding in the presence of the fungicides and osmotic sensitivity, suggesting a connection between fungicide mode of action and morphogenesis and glycerol accumulation (which would also be related to osmotic, turgor regulation and osmotic shock), which may explain why this mutation has never been reported from field isolates with fungicide resistance phenotypes (Ramesh et al. 2001).

Studies by Leroux et al. (1999) in *B. cinerea* in French vineyards found strains resistant to multi-site fungicides, and multiple combinations of fungicide resistant phenotypes in vineyards in France: two types of benzimidazole (e.g. carbendazim, thiabendazole)-resistant strains were detected, with negative cross-resistance towards the n-phenylcarbamates (diethofencarb, diphenylamine and dicloran) found only in one type of resistant strain, as previously stated above. However, this study also found that most dicarboximide (e.g. iprodione, procymidone, vinclozolin)-resistant strains were also weakly resistant to aromatic hydrocarbon fungicides (e.g. chloroneb, dicloran, tolclofos-methyl) but remained sensitive to phenylpyrroles (e.g. fenpiclonil, fludioxonil). However, in some other dicarboximide-resistant strains, resistance was observed either as being restricted to dicarboximides or as extending weakly to phenylpyrroles. Dicarboximides, phenylpyrroles and the aromatic hydrocarbon fungicides (e.g. chloroneb (PCNB), dicloran, quintozene, tolclofos-methyl) are not chemically related, but obvious structural similarities can be observed (Fig. 3), and in laboratory studies, mutants of *B. fuckeliana* and several other fungi, have a positive cross-resistance between them (Leroux et al., 1992;Faretra and Pollastro, 1993).

Similar observations regarding the complexity of resistance profiles were observed in *B. cinerea* in the laboratory (Oshima et al. 2002). Dicarboximide- resistant laboratory mutants selected for high resistance to dicarboximides, aromatic hydrocarbons, and phenylpyrroles, were hypersensitive to osmotic stress, and were rarely obtained from the field (Leroux et al.,

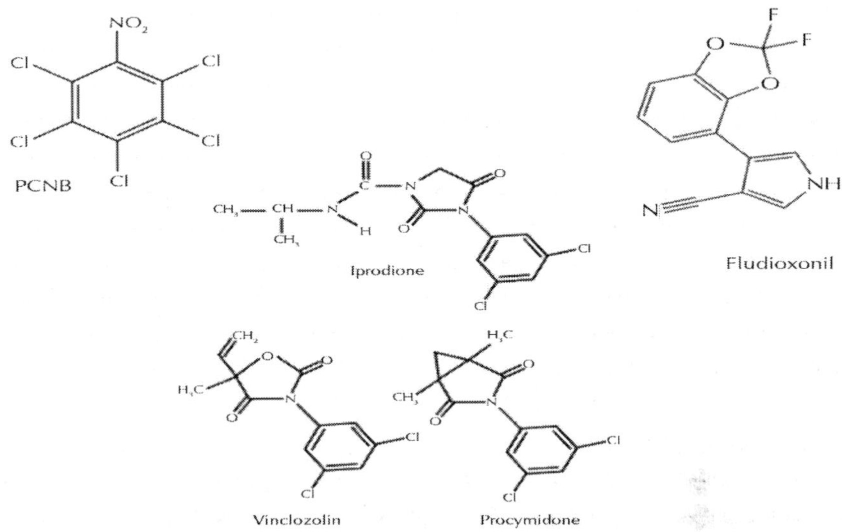

Figure 3: Comparison of members of the aromatic hydrocarbons (PCNB), phenylpyrrole fungicide fludioxonil and three dicarboximides, iprodione, vinclozolin and procymidone. All five structures share low molecular weight and aromatic hydrocarbon moieties that may explain why some cross resistance between these distinct classes of fungicides is observed.

1992; Faretra and Pollastro, 1993; Lyr, 1995). Instead, only moderately resistant, osmotically-stable strains were recovered from the field, and these field isolates of *B. cinerea* showed cross-resistance to aromatic hydrocarbons but not to phenylpyrroles (Oshima et al. 2002). In fact, phenylpyrrole- resistant mutants from *B. cinerea* that were first isolated and characterized in the laboratory were due to mutations at a different locus (*Daf1*—later identified as *BcOS1*) locus responsible for dicarboximide resistance (Faretra and Pollastro, 1993) and were only later found in the field, and were only described as "weakly resistant"(Leroux et al. 1999). These early studies found that mutations of the *Dafl* gene are highly variable, with at least five classes of alleles identified and associated with differing levels of resistance (sensitive (S), low resistance (LR), moderate resistance (MR) or highly-resistant (HR)) to dicarboximides and phenylpyrroles, respectively: S/S, LR/S, HR/LR, HR/MR, HR/HR (Faretra and Pollastro, 1993). Field resistance was due to a single amino acid substitution, from Ile to Ser, which occurred at codon 365 in *BcOS1* gene (Oshima et al. 2002), and at two other amino acid substitutions at the same position, 365 Ile to Asn or Arg, in field-resistant isolates. Work done by Oshima et al. (2006) used PCR-CAPS, exploiting a *Taq*1 restriction site that results in the mutant allele that confers resistance. Approximately 41 percent of the isolates had the type I mutation that could be detected by PCR-CAPS. However, other types of dicarboximide resistance were detected and sequence analysis of these mutants classified them into type II isolates that have three amino acid substitutions within*BcOS1*p (V368F, Q369H, and T447S) or type III isolates that have two amino acid substitutions within *BcOS1*p (Q369P and N373S) (Oshima et al. 2006).

Many fungi accumulate glycerol to increase osmotic pressure to contend with osmotic stress, or for pathogenicity. Studies done by Fujimura et al., (2000) in *Neurospora crassa* with osmotic-sensitive (os) mutants with *os-1, os-2, os-4* and *os-5* mutations (similar to the previously described resistances in*B. cinerea*) showed cross-resistance to dicarboximides (iprodione and vinclozolin), but also the aromatic hydrocarbons (PCNB). All of the os mutants except for some os-1 mutant alleles were resistant to the phenylpyrrole fungicide, fludioxonil (Ochiai et al. 2002). Previous work by Grindle (1982) found other osmotic sensitive mutants (cut) that were not resistant to these fungicides, suggesting that osmotic sensitivity alone was not the mechanism. However, the original report byGrindle (1982) on dicarboximide resistance in *N. crassa* shows that multiple alleles that segregate independently were involved with this os-1 resistance, suggesting that other genes may still be involved in resistance (See section on MDR for further information). Work by Zhang et al. (2002), identified a mitogen activated protein (MAP) kinase (HOG1) that was resistant to phenylpyrrole, and were caused by a frame shift from tryptophan to a stop codon or nonsense point mutations, allowing for a screenable genotype. However, field resistance of phenylpyrrole has not been reported to date, and the complex phenotype involved in this type of resistance makes it biologically interesting, but hardly defining in the diagnosis of fungicide resistance.

Sterol Biosynthesis Inhibitors

This group of fungicides exploits one of, if not the most important mode of action for both antifungals in medicine and fungicides in agriculture. The sterol biosynthesis inhibitor (SBI) fungicides inhibit a precursor of ergosterol that is essential for the development of the fungal membrane (Brent, 1995). These fungicides can be classified based upon their target sites in sterol biosynthesis, with inhibitors of squalene epoxidase (e.g. naftifine, terbinafine, tolnaftate) primarily used for mammalian mycoses. The remaining classes are used for agricultural purposes. 'Amines' or 'morpholines' (e.g. fenpropidine, piperalin, spiroxamine, tridemorph) act as inhibitors of sterol Δ14-reductase or Δ8->Δ7-isomerase, and target the products from the *Erg24* and *Erg2* genes; hydroxyanilides, represented by fenhexamid, act on the 3-keto reductase, C4- demethylation encoded by *Erg27* and is specifically used for the control of *Botrytis;* inhibitors of sterol 14-alpha-demethylase (e.g. bitertanol, triazoles, imidazoles, in addition to imazalil, prochloraz, pyrifenox, triadimenol) are referred to as demethylase inhibitors (DMIs) and primarily target *Erg 11/CYP51*(FRAC 2012), and are effective against a wide variety of phytopathogens.

DMIs were introduced for plant disease management in the 1970s, with resistance and reduced efficacy reported soon after (Brent, 1995). Due to both agricultural and medical importance, the molecular mechanisms of SBI resistance have been highly studied and include (i) mutations in 14-alpha-demethylase (CYP51) structural gene (Delye et al. 1997; 1998; Canas-Gutierrez et al. 2009, and others) and ERG27 gene (Albertini and Leroux, 2004; Fraaije et al. 2007); (ii) overexpression of the CYP51 gene, leading to increased production of the target enzyme (Cools et al. 2012; Luo et al. 2008;Ma et al. 2006; Schnabel and Jones, 2001; Hamamoto et al. 2000), and (iii) overexpression of the ATP-binding cassette (ABC) transporters (Zwiers et al. 2002; Hamamoto et al. 2000; Nakaune et al. 1998) which will be addressed later in the chapter. Several studies, at least in the laboratory, have demonstrated that multiple mechanisms contribute to the variation in azole susceptibility (Stergiopoulis et al. 2003; Zwiers et al. 2002), and genetic analysis of progeny from a cross between *M. graminicola*isolates with differing sensitivities to DMI fungicides revealed a continuous distribution of resistance, leading the researchers to conclude on the polygenic nature of inheritance (Stergiopoulis et al. 2003), leaving open the possibility that other mechanisms may also be at work.

Mutations in the 14-alpha-demethylase (*ERG11/CYP51*) structural gene that lead to a decreased affinity of the target protein to DMI fungicides have been found in powdery mildews (Delye et al. 1998;

Delye et al. 1997); two *Mycosphaerella* pathogens—*M. graminicola* (Leroux et al. 2007; Cools et al. 2006) and the black Sigatoka fungus, *M. fijiensis*, (Gutierrez-Canas et al. 2009); and the cereal leaf spot pathogens--*Tapesia* spp. (Albertini et al. 2003), although work by Wood et al. (2001) did not find a correlation between resistance and the presence of known mutations in the structural gene. In *M. graminola*, reduced sensitivity to DMIs could be correlated with an alteration in the CYP51 structural gene at codons 459–461 (Cools et al. 2005; Leroux et al. 2007). In the barley powdery mildew,*Blumeria graminis*, amino acid substitutions in CYP51 at Y136F and K147Q, were detected and a very high level of resistance was associated with the allele containing K147Q mutation. Sequence analysis of the CYP51 gene from the progeny of a cross between DMI-sensitive and resistant isolates demonstrated co-segregation between the mutant alleles and resistance. Consistent with other studies, the authors found that genetic analysis of resistance to the triadimenol indicates that mutation of the*CYP51* gene is not the only mechanism of resistance operating in *B. graminis*: Two moderately resistant isolates had no mutations in the *CYP51* gene, and had identical sequences to that of the sensitive isolate, suggesting that resistance in at least these two isolates must be due to a mutation in an entirely different gene (Wyand and Brown 2005). The identification of isolates such as these should serve as adequate warning that molecular detection of mutations in the structural gene may result in an under-reporting of actual incidence of resistance in the field. Finally, mutations in the structural gene of *ERG27* in *B. cinerea* were detected and correlated with mutations (F412I and R496H) that would lead to a decreased affinity of fungicide to the target protein, but it was regarded as low resistance or moderate resistance (Albertini and Leroux, 2004; Leroux et al. 2007).

A second mechanism of DMI resistance involves overexpression of the *CYP51* gene. Overexpression of CYP51 has been associated with DMI resistance in *P. digitatum* (Hamamoto et al. 2000), *V. inaequalis*(Schnabel and Jones, 2001), *M. fructicola* (Luo et al. 2008), *B. jaapi*i (Ma et al. 2006), C. *beticola*(Bolton et al. 2012) and *M. graminicola* (Cools et al. 2012). In all but *C. beticola*, overexpression of CYP51 has been associated with insertions in the upstream promoter: A tandemly repeated 126 bp fragment was found in the promoter of *P. digitatum*; Ma et al. (2006) identified a variably-sized retrotransposon-like element in *B. jaapi*; Schnabel and Jones (2001) reported on a 533-

bp fragment found in some (but not all) DMI-resistant isolates of *V. inaequalis*; and Luo et al. (2008) found a 65 bp 'Mona' element in the promoter of *CYP51* in *M. fructicola* that was strongly linked to the DMI resistance phenotype. However, subsequent work by Villani and Cox (2011), found that the 'Mona' element was present in a range of sensitivities, and not always present in resistant isolates, suggesting it is just one of many possible mechanisms of resistance. Overexpression of *ERG27* has not yet been identified as having a role in resistance to the hydroxyanilide class (FRAC 17) of fungicides.

A final mechanism of DMI resistance relies upon "drug" transporters, trans-membrane proteins located in the plasma membranes that utilize ATP to translocate compounds, including toxins, out of the cell, preventing the accumulation of these products to toxic levels, as has been shown for *B. cinerea*(Leroux and Walker 2011; Leroux et al. 2002), *M. graminicola* (Cools et al. 2007) *Penicillium digitatum* (Nakaune et al. 1998), and possibly *V. inaequalis* (Koller and Wilcox, 2001). This will be discussed in a later section. Some moderate resistance to DMI fungicides in *B. cinerea* (termed Ani R2 and Ani R3 by Leroux et al., 1999) was previously described (Stehmann and DeWaard, 1995; Del Sorbo et al. 1997). Later studies found isolates with this similar phenotype demonstrated pronounced increases in expression level of the ABC transporter genes (Kretschmer et al. 2009) and not the structural genes themselves, creating greater complexity and less certainty in molecular detection of fungicide resistance. As a result, DMI resistance in numerous species has been identified, but only 10 species have had the molecular mechanism of resistance studied in depth.

Although fungicide resistance baselines and thresholds have been previously established in many agricultural systems, the primary stumbling block to the rapid identification of DMI fungicide resistance for most fungi is the nature of its quantitative resistance (Koller et al. 1991; Schnabel and Jones, 2001; Ishii, 2009). Unlike qualitative resistance that results from a single gene mutation (e.g.,*TUB1R* conferring resistance to benzimidazoles), DMI resistance is much more complex. Due to the diversity of mechanisms of DMI resistance in different pathogens (and even the variation in response by pathogens to different DMIs), it is unlikely that any one approach, with the exception of phenotypic analysis, will provide the screening necessary to detect fungicide resistance in a diversity of pathosystems, although the possibility exists that the monogenic DMI resistance observed in powdery mildew pathogens may be amenable to this approach. In V. *inaequalis*, examples of DMI resistance to date are associated with an insertion in the promoter of the CYP51 target gene that has been correlated with overexpression of CYP51. The authors of this work (Schnabel and Jones, 2001) noted that the correlation was strong, but not absolute. Subsequent work to utilize this knowledge to develop PCR based screening of resistant isolates can detect the insertion, but not predict the degree of resistance that results (Villani, Cox and Beckerman, in preparation and Fig 4.), preventing the rapid detection of resistance in the field. Ultimately, *in vitro* screening of radial growth in the presence of fungicide (a 4-week process after a single-spore isolate in pure culture is established) is needed to determine the degree of resistance in the isolate, and at least in *V. inaequalis,* it does not seem to be currently possible to rapidly detect fungicide resistance in the field, and certainly not to a degree that thresholds could be identified to warn when a fungicide failure will occur. This situation is quite different than the detection of fungicides that impact respiration.

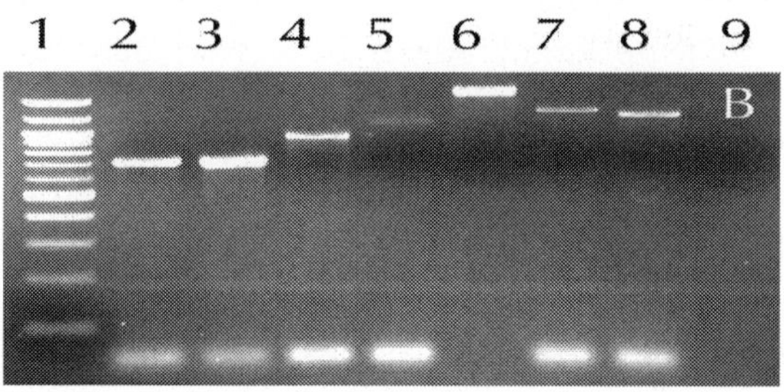

Figure 4: Amplicon size is not related to overexpression of *CYP51* in *V. inaequalis*. Primers flanking the promoter region of *CYP51* were used to examine the relationship between the 533 bp insertion, size of promoter insertion and resistance to myclobutanil. Many of the highly resistant isolates did not yield an amplicon, and those that did did not have the expected 533 bp insertion. From Villani, Cox, and Beckerman, unpublished.

The Respiration Inhibitors

Respiration is an obvious target for fungicides. The oldest synthetic fungicides still used today include dithiocarbamates (e.g., mancozeb, thiram, ziram) or pthalimide derivatives (e.g., captan), and prevent plant disease by the inhibition of spore germination, and subsequent germ tube elongation. To date, there have been no credible reports of resistance to these older fungicides. Furthermore, older synthetic multisite fungicides are effective against a wide range of plant-pathogenic fungi, and Chromista. These multisite inhibitors are thio-reactant with those enzymes involved in respiration (Lyr, 1977); Other inhibitors of respiration work at complex II at the succinate dehydrogenase gene, complex III, the cytochrome bc1 (ubiquinol oxidase) at Qo site, or cytochrome bc1 (ubiquinol oxidase) at Qo site; or complex V, where fungicides like fluazinam actually work by uncoupling electron transport and ADP phosphorylation of oxidative phosphorylation (FRAC, 2009). As such, this target is essential to aerobic eukaryotes, and present targets for disruption. The most recently developed group of respiration inhibitors targets mitochondrial complex III (the cytochrome bc1 complex), and targets the outer quinol-oxidizing pocket (Qo site), and complex II at the succinate dehydrogenase gene, which are the areas of most recent and active interest due to their recent release, the nature of fungicide resistance, and potential ease of detection.

QOI—the Quinone Outside Inhibitors

Fungicides that inhibit the cytochrome bc1 enzyme complex (complex III) at the Qo site (Qo inhibitors, QoIs) were first introduced to the market in 1996. QoI resistance has now been reported in more than 20 pathogens, including the Chromista pathogens, downy mildew of grape and cucurbit, and the fungi that cause apple scab, wheat powdery mildew and Septoria leaf blotch, to name but a few. Resistance has been reported in all major continents: Asia, Australia, Europe, North America, and South America (FRAC 2006; Wilson and Wicks, 2011). The majority of the reports associate the point mutation that results in a change from glutamate to alanine at codon 143 (G143A), followed by the resistance that results from a transition from phenylalanine to leucine at position 129 (F129L) (Pasche et al. 2005; Sierotzki et al. 2007); the least commonly observed mutation involves a change from glycine to arginine at position 137 (G137R) (Sierotzki et al. 2007), and of course, unknown causes (Lesniak et al. 2011). These easily identified mutations allow for the rapid detection for the potential of QoI-resistance in a variety of fungi and Chromista. Furthermore, the degree of resistance is much greater in the G143A allele (referred to as high resistance), and has consistently resulted in a significant reduction in disease control, as compared to resistance associated by F129L and G137R which results in more moderate or partial resistance, and some degree of control at highest rates of fungicide.

A more recent study by Lesniak et al. (2011), found ninety-eight percent of QoI-resistant/shifted isolates screened in Michigan were associated with G143A, consistent with other phytopathogenic fungi resistant to this class of fungicides (Ishii et al., 2009; Ishii et al. 2007). Sequencing the entire*CYTb* gene of resistant isolates that did not test positive for the G143A transition did not reveal other mutations in the structural gene, and suggests other mechanisms of resistance (Lesniak et al. 2011). Regardless of mutation present, functional analysis of the *CYTb* gene is confounded by the complexity of fungal mitochondrial genetics. A fungal cell may contain >250 mitochondria and an equal copy number of mitochondrial DNA (mtDNA). A single point mutation in a single copy of a mitochondrial gene could be detected, but would not be sufficient to confer resistance, unless multiple copies of the allele were present. However, over time, and under increasing selection pressure, the allele would become fixed, and the mitochondria population within the cell would shift from sensitive to increasing numbers bearing the resistant allele, shifting resistance and becoming homeoplasmic over time. Thus, early detection of this shift could preserve long-term fungicide efficacy by short-term use of alternative fungicides until the allele shifts back to the previous susceptible homeoplasmy. Early detection with quantitative PCR (qPCR) could theoretically enable the assessment of either gene expression or copy number that is required for QoI resistance to develop the field. However, to date, this work has not been attempted with any field populations. Preliminary data suggests that the populations in New York, Michigan and Indiana are not yet fixed (heteroplasmy). Laboratory analysis of QoI resistant mutants found that serial passage of trifloxystrobin-resistant isolates in vitro has resulted in a loss of resistance In *V. inaequalis*,after only two rounds of propagation allowed a significant majority of mitochondria containing the wild-type cytochrome b sequence to re-appear; This was not observed when the fungus was under continued selection (Zheng et al. 2000). The conversion of heteroplasmic isolates back to wild type in the absence of QoI fungicides would imply that a G143A mutation is associated with a fitness penalty, although Chapman et al. (2011) found no evidence of this. Early detection of the resistance allele, and prophylactic switching to non-QoI fungicides may return the population of a sensitive level in a few years, and with proper monitoring, allow the use of QoI fungicides for scab control again. Similar findings on *M. graminicola* also showed conversion of mutated *CytB* gene back to wild-type in *M. graminicola* (Fraaije et al. 2002).

With QoI resistance, allele specific PCR (Lesniak et al. 2011 and many others) and restriction enzyme digest of the amplicon (Leroux et al. 2010 and many others) are both used to detect the presence of the allele. Work by Fontaine et al. (2009), found that the QoI-resistant allele could not be detected after*Fnu*4HI digestion for R : S ratios equal to or below 1:9 (w:w). When compared to the allele specific (AS-PCR) PCR, the mutant allele was amplified at the lowest ratio tested, much more sensitive than the PCR-CAPS technique (Fontaine et al. 2009). However, work by Chapman et al. (2011), suggests an over prediction of resistance, if the mutation is used as a sole criterion for identification of resistance. Similar finding occurred in work done by Lesemann et al. (2007), on apple powdery mildew, with PCR of cleaved amplified polymorphic sequences (CAPS) analyses suggest that the proportion of mitochondria carrying the G143A exchange determines the degree of strobilurin resistance. Lesemann et al. (2007) also found a high variability in the cytB gene of *P.leucotricha* and discusses the role that mitochondrial heteroplasmy may play in conferring a selective advantage under changing conditions, as does Avila-Adame et al. (2003). Many papers on QoI-resistance in phytopathogens fail to discuss mitochondrial heteroplasmy (Patel et al. 2011). This is important as for*P. leucotricha* (Lesemann et al. 2007) and *V. inaequalis* in the laboratory (Zheng et al.2000) and in the field (Lesniak et al. 2011), clearly demonstrated that all isolates of the apple powdery mildew and scab pathogens tested were heteroplasmic for the G143A mutation. Lastly, a study by Miguez et al. (2003), examined the role of alternative oxidase in reducing the sensitivity of *M. graminicola* to the QoI azoxystrobin. Although the level of resistance was lower than the more commonly identified, G143A, sufficient resistance was incurred despite the absence of this mutation, suggesting yet another mechanisms of resistance.

SDHI

Succinate dehydrogenase inhibitor (SDHI) fungicides include "first generation" SDHIs, like flutolanil, carboxin, followed by later (and improved) "second generation" compounds, including boscalid, and even "third generation" products like penthiopyrad, and fluopyram. This class of compounds targets and binds to the ubiquinone-binding site (Q-site) of the mitochondrial complex II, specifically the succinate dehydrogenase (SDH) complex in the respiratory chain also referred to as complex II or succinate:ubiquinone oxidoreductase (SQR), inhibiting fungal respiration by blocking electron transport [Kuhn, 1984; reviewed by Avenot and Michailides (2010), including a recent list of resistant organisms and corresponding mutations]. There is no evidence of cross-resistance with other similarly acting chemical classes such as QoI fungicides (which also affect energy production and electron transport) due to their unique mode of action and target site. The primary target of the SDHI fungicides is the SDH complex that consists of four subunits: a flavoprotein (Fp) subunit (SdhA), an iron-sulfur protein (Ip) subunit (SdhB), and two membrane-anchored protein subunits (SdhC and SdhD). Mutations conferring resistance have been found in the SDHB (Avenot et al 2008) and SDHC (Ito et al. 2004) and SDHD subunits (Avenot et al. 2009).

The first widely used SDHI fungicide, carboxin, was used in the late 1960s, and its applications were limited due to its primary activity against basidiomycete pathogens, namely rusts, *Rhizoctonia* spp., and corn smut (*Ustilago maydis*), and limited activity against other pathogens (Sisler 1988). In contrast to old SDHIs, newer active ingredients of SDHIs comprise compounds such as boscalid, penthiopyrad or fluopyram, and are characterized by a broad spectrum of fungal activity on various crops (Stammler et al., 2007; Stammler et al., 2006) particularly *Botrytis* and *Alternaria* species. Carboxin is less efficient than boscalid in controlling the wild-type sensitive isolates (Avenot et al. 2008). Unlike carboxin, boscalid prevented *B. cinerea* spore germination completely at high concentrations, and inhibited germ-tube elongation at low concentrations. This phenomenon was particularly noticeable if biological tests were conducted in media containing succinate rather than glucose as the carbon source (Lyr, 1977). This is an important point when performing in vitro screens of fungicide resistance phenotypes.

Field and laboratory mutants resistant to carboxin and boscalid have been reported in a variety of fungi and cropping systems (For a recent review see Avenot and Michailides 2010). Carboxin and other SDHI fungicide resistances are described as monogenic, and have been identified in both field and laboratory. Sequence analysis of the gene encoding the target protein, the succinate dehydrogenase enzyme (SDHB), revealed that single or double point mutations in the highly conserved regions of gene were associated with resistance. Different levels of resistance are associated with mutations in different alleles: In *B. cinerea,* mutations resulted in P225L or P225F transitions that confer high resistance, and in a histidine to tyrosine replacement at position 272 (H272Y) or Arg (H272R) (Avenot et al. 2008), although Angelini et al. (2010) found these two mutations resulted in a lower level of resistance. In *A. alternata,* sequence analysis of the SDHB gene from sensitive and resistant isolates identified a H277Y and H277R transitions, as well. (Avenot et al. 2008). It is important to note that in some boscalid-resistant strains of *A. alternata,* there were no identifiable mutations in the AaSdhB gene as compared to the wild type, suggesting that mutation(s) in other loci are involved in the boscalid resistance phenotype. This is not surprising since the boscalid mode of action involves at least two other genes of the Sdh complex (SdhC or SdhD) (Avenot et al. 2009).

Despite the elucidation of the molecular mechanisms of resistance, mycelial growth assay in liquid medium in microtiter plates were developed to monitor fungicide resistance, in *A. alternata* (Avenot and Michailides 2007) and a variety of other fungi, eventually resulting in the development of rapid in vitro monitoring procedures with a single discriminatory dose of boscalid (Avenot and Michailides, 2010), despite the eventual elucidation of many of the molecular mechanisms of resistance.

MDR-Efflux Pumps and Resistance

All living organisms, eukaryote and prokaryotic, are exposed to both food and toxins in their natural environments. To quote Pao (1998), "Transport systems allow the uptake of essential nutrients and ions, excretion of end products of metabolism and deleterious substances, and communication between cells and the environment." Some of these deleterious substances may be antibiotics produced by bacteria and actinomycetes, antifungals [e.g., strobilurin A produced by *Strobiluris tenecellus*, killer yeast strains, plant defense compounds (phytoalexins, alkaloids, small molecular weight peptides)], and heavy metals. There are two protein families that are key players in this type of transport process: The ATP-binding cassette (ABC) and the major facilitator superfamily (MFS) of transporters. ABC transporters are able to bind and hydrolyze nucleotide triphosphates (mainly ATP) due to a conserved cytosolic, nucleotide-binding fold (NBF or ATP-binding domain) and use this energy to transport solutes across cell membranes (Higgins, 1992). MFS transporters work as a "secondary" active transport system that does not require ATP for functionality, and are capable only of transporting small solutes in response to chemiosmotic ion gradients (Pao 1998). These two families of transporter proteins can mediate a quantitative multidrug resistance (MDR) to multiple classes of fungicides, however, the resistance levels conferred against individual fungicides are greatly reduced as compared to fungicide resistance conferred by target site mutations (de Waard et al. 2006; Mernke et al. 2011).

In the human pathogen *C. albicans*, the role of MDR transporters and resistance is fairly straightforward: The ABC transporter, CDR1, plays a major role in resistance (in a majority of isolates) to fluconazole and miconazole due to upregulation of the gene (Sanglard et al., 1995) resulting in increased efflux of the antifungal and its decreased accumulation in the cell, thereby reducing inhibition of the ERG11/CYP51 gene in most isolates resistant to fluconazole. In addition to its role in resistance to azoles, CDR1 also confers resistance to other sterol biosynthesis inhibitors, including allylamines and morpholines, in addition to several other drugs. A separate transporter, referred to BENR, was highly overexpressed and conferred resistance to azoles and benomyl (Sanglard et al., 1999; Sanglard et al., 1995).

Detection of this type of fungicide resistance has proven to be more difficult, or at least less published, in plant pathogens. Numerous reports exist about multiple fungicide resistances in plant pathogens (Leroux et al. 1997, Nakaune et al. 1998; Leroux et al. 1999; Kretchmer et al. 2011; Chapman et al. 2011). It is important to stress that not all of these multiple fungicide resistances are due to MDR, but in fact, are due to other mechanisms (see the section on Dicarboximides for elucidation of one such mechanism) including segregation of separate, multiple resistances (Chapman et al. 2011). One of the first studies (Leroux et al. 1999) that screened multiple fungicide resistances in *B. cinerea* found resistance to anilopyrimidines (AP, e.g. cyprodinil, mepanipyrim, pyrimethanil) and identified two distinct resistance phenotypes: The most AP-resistant isolates were resistant to only anilinopyrimidines, suggesting action on the hitherto unidentified, single target gene (Hilber and Hilber-Bodmer 1998) and a second phenotype that included resistance to distinctly different classes of fungicides including dicarboximides, phenylpyrroles, sterol biosynthesis inhibitors (e.g. tolfanate, prochloraz, tebuconazole) and the hydroxyanilide derivative, fenhexamid, consistent with an MDR phenotype.

The role of MDR in fungicide resistance was further strengthened by work by Kretschmer et al. (2011) who found that fungicide resistant field isolates from France and Germany exhibited three distinct MDR phenotypes of increased fungicide efflux activity and overexpression of efflux transporter genes. In this study, MDR1 strains were found to possess mutations in the transcription factor, *Mrr1*, that controls the ABC transporter gene AtrB; the MDR2 strains possessed insertions of a retrotransposon-derived sequence in the promoter region of the major facilitator superfamily (MFS) transporter gene mfsM2 (more thoroughly described by Mernke et al. 2011), and the MDR3 strains which showed the highest levels and broadest spectrum of resistance against most fungicides tested, and was identified as recombinants carrying both MDR1-specific mutations in mrr1 and MDR2-specific mutations in mfsM2 (Kretschmer et al. 2011). All MDR strains showed strong constitutive overexpression of either one (MDR1, MDR2) or two (MDR3) drug efflux transporter genes. MDR1 and MDR3-described isolates had an increased efflux for fludioxonil whereas MDR2 strains did not, while bitertanol efflux was observed for all MDR phenotypes, although MDR1 possessed a less resistant phenotype. Prior to this, Hayashi et al. (2001) found in laboratory isolates, overexpression of the ABC Transporter Gene*BcatrD* involved in resistance to two azoles, in addition to the dicarboximide fungicide iprodione, the benzimidazole fungicide carbendazim, and the antibiotic cycloheximide.

In *M. graminicola*, laboratory isolates were selected with decreased azole susceptibilities and cross-resistance to chemically unrelated of low molecular weight compounds (Zwiers et al. 2002). Later studies found field isolates showed differences in both basal and induced levels of ABC transporter gene transcript, although no correlation between increased expression and azole sensitivity was evident (Stergiopoulos et al., 2003); Cools et al. (2007) were unable to establish a direct relationship between over-expression of the ABC transporters and decreased azole sensitivity, and later studies, using cDNA microarrays to profile the transcriptional response of *M. graminicola* to epoxiconazole, and compared the expression profiles of an azole-sensitive and less sensitive *M. gramincola* isolates did not find a relationship. They found upregulation of ten genes that provided different constitutive expression profiles between the two strains, including drug transporters, a cell surface glycoprotein, stress response protein rds1, and an unknown gene encoding a homologue of the antibiotic response protein in addition to differential expression between components of the sterol biosynthesis pathway between sensitive and less sensitive isolates, and components of the mitochondrial respiratory chain. (Cools et al. 2007). Thus, at least in *M. graminicola*, studies have not demonstrated a relationship between expression (or overexpression) of ABC transporter genes directly (Stergiopoulos et al. 2003) or by microarray (Cools et al. 2007), despite its demonstrated role in field resistance in *B. cinerea*.

Although the role of MDR genes in fungicide resistance is clear in some pathogens, rapid detection of fungicide resistance due to MDR is not. Some of these difficulties reside in the unclear and multiple mechanisms of DMI and other fungicide resistances discussed above, others in the role and regulation of MDR. To date, most differences identified in fungicide resistance where MDR is implicated have been found to be due to overexpression of ABC transporter genes. Most important to stress is the lack of correlation between expression level specific ABC transporter gene with fungicide resistance, suggesting that multiple transporters may be involved (at least in DMI resistance) or that other mechanisms in addition to upstream transcription factors, have not yet been identified (Cools et al. 2007; Leroux et al. 2011).

CONCLUSIONS

Fungicide resistance can be conferred by a variety of mechanisms, and plant pathologists need to recognize that multiple genes may be necessary for this resistance, and that single genes, although easily scored or detected by molecular means, may not be sufficient for determination of resistance. Central to these studies is the recognition of the strengths and the limitations of molecular detection, and the importance of phenotypic versus genotypic resistance. This requires that we recognize that the phenotype of fungicide resistance consists (in many instances) of "major" genes, "minor" genes, the interactions between these genes (epistasis) and the interactions between genes and the environment, and that many of the known mechanisms conferring quantitative fungicide resistance utilize alternative metabolic pathways, exclusion or efflux of fungicides, and hitherto unknown mechanisms (Brent and Holleman 1998; Leroux et al. 1999; Lesniak et al. 2011).

The detection of single gene targets is only the beginning, and understanding the mechanism of resistance does not provide a blueprint as to how to manage the crop in the absence of the fungicide lost. Previous studies labored under an assumption that "a timely detection of resistance levels in populations of phytopathogenic fungi in a field would help growers make proper decisions on resistance management programs to control plant diseases" (Ma and Micheilides 2005). It certainly doesn't hurt, but few instances are available where this knowledge has been translated for use to growers and resulted in in-season changes of management beyond cessation of use of the fungicide in question. To date, most resistance management consists of abandoning the fungicide that has failed, to use others (Chapman et al. (2011); Lesniak et al. (2011), Avenot et al., others); Chapman et al. (2011) is one of the few instances where fungicide resistance detection was used many years later to confirm that a fungicide could be used with some degree of successful certainty. This was determined by mycelial growth assay as no molecular detection methods currently exist to screen for dodine resistance.

Fungicide sensitivity testing using mycelial growth or germination inhibition (in petri dishes or microtitre plates) lacks the excitement and appeal of the many different molecular approaches to detect fungicide resistance. It is certainly not as fast as any molecular test, although microtiter-based assays combined with Alamar blue (AB), or resazurin provide quantifiable and early detection within days, for fast growing pathogens like *Monilinia* (Cox et al. 2009) or *Verticillium* (Rampersad 2011). AB is an oxidation-reduction indicator dye used to detect microbial respiration. In the presence of actively growing cells, the resazurin indicator is changed from an oxidized, nonfluorescent blue form to a reduced, fluorescent pink form. Inhibition of growth maintains an oxidized environment, leaving the indicator blue. Results can be easily discerned with the naked eye due to the colorimetric nature of the test, or more rapid and sensitive measurements can be taken with spectroscopy equipment. In these studies, the AB assay provided a rapid and reproducible method of testing fungicide efficacy and provide an option of deriving quantitative data in the form of degree of resistance versus the qualitative data associated with molecular detection. Use of AB in fungicide resistance requires aseptic techniques that are essential for any microbial assay (Cox et al. 2009). This prerequisite proved to be insurmountable for use of this technique for fungicide resistance screening in *V. inaequalis*, and possibly other fungi. Finally, the calculated the cost of 96 reactions, excluding labor and plate reader, was found to be under $4 per isolate for AB assays and over $15 per isolate for mycelial growth assays (Cox et al. 2009) making this affordable for multiple labs and diagnostic clinics to screen for fungicide resistance as a service. Despite the lack of intellectual appeal, there are other significant advantages of fungicide resistance phenotype screening, chief amongst which is the low cost, accuracy and reliability of the screening for the fungicide resistant phenotype over the genotype, because, unless all known genes and alleles are screened are known and screened, the risk of under-reporting actual resistances exists (Quello et al. 2010), and is rarely discussed.

This is not to say that molecular detection of fungicide resistance is without merit. One of, if not the most critical challenges facing applied plant pathologists is the need for the early detection of fungicide resistance in a population. Molecular methods can be used to powerfully detect and monitor the emergence of resistance in those instances where detection is possible, to a degree that phenotypic assays cannot achieve. Unfortunately, there are few studies where this has been done, and most research ends with a proof of concept, and not the implementation of the developed technique to better manage fungicide resistance. Currently, there are few studies monitoring sensitivity and early detection through the use of sentinel plants coupled with molecular detection in those instances where the mechanism can be detected, which could provide powerful information in an early warning system to anticipate the emergence of fungicide resistance in a population and prevent its occurrence. One problem with this approach is that most models for the detection of fungicide resistance emergence should assume an initial resistance mutation frequency from anywhere from as high as 10^{-4} (Zwiers et al. 2002) to as low as 10^{-8} (Zheng et al. 2000) based upon laboratory studies.

The developing paradigm that has emerged has been the generation of mutants in the laboratory, followed by their isolation from the field in some instances (but not all), followed by phenotypic screen, and identification of mechanism (Albertini et al. 1999 to Zwiers 2002, and others.) Genotypic screens have been developed, but are rarely evaluated for efficacy and accuracy over time, let alone predictive ability or utility. Furthermore, these developments have still not helped the farmer, nor provided him or her with the information necessary to make the appropriate in-season management decision. The question that remains is if we should be willing to shift the paradigm and use the best technique to adequately address the problem of fungicide failure in the field. As we enter a post-genome period, the technology is available for the detection of even single nucleotide polymorphisms. Whether it is available to rapidly detect fungicide resistance in plant pathogens in a way that is useful for growers remains to be seen.

ACKNOWLEDGEMENT

I would like to thank Drs. Rick Latin and Kiersten Wise for our numerous and ongoing discussions on fungicide resistance, and for their review of this manuscript.

REFERENCES

1. C. Albertini, M. Gredt, and P. Leroux, 1999Mutations of the b-tubulin gene associated with different phenotypes of benzimidazole resistance in the cereal eyespot fungi Tapesia yallundae and Tapesia acutormis. Pestic. Biochem. Physiol. 641723
2. C. Albertini, M. Gredt, and P. Leroux, 2003Polymorphism of 14 alpha-demethylase gene (CYP51) in the cereal eyespot fungi Tapesia acuformis and Tapesia yallundae. Eur. J. Plant Pathol. 109117128
3. C. Albertini, and P. Leroux, 2004A Botrytis cinerea putative 3-keto reductase gene (ERG27) that is homologous to the mammalian 17

beta-hydroxysteroid dehydrogenase type 7 gene (17 beta-HSD7). Eur. J. Plant Pathol. 110723733

4. H. F. Avenot, andT. J. Michailides, 2010Progress in understanding molecular mechanisms and evolution of resistance to succinate dehydrogenase inhibiting (SDHI) fungicides in phytopathogenic fungiCrop Prot. 29643651
5. H. F. Avenot, A. Sellam, andT. J. Michailides, 2009Characterization of mutations in the membrane-anchored subunits AaSDHC and AaSDHD of succinate dehydrogenase from Alternaria alternata isolates conferring field resistance to the fungicide boscalidPlant Pathol. 5811341143
6. H. F. Avenot, A. Sellam, G. Karaoglanidis, and T. J. Michailides, 2008Characterization of mutations in the iron-sulphur subunit of succinate dehydrogenase correlating with boscalid resistance in Alternaria alternata from California pistachio. Phytopathology98736742
7. C. Avila-adame, G. Olaya, and W. Köller, 2003Characterization of Colletotrichum graminicola isolates resistant to strobilurin-related QoI fungicidesPlant Dis. 8714261432
8. S. Banno, F. Fukumori, A. Ichiishi, K. Okada, H. Uekusa, M. Kimura, and M. Fujimura, 2008Genotyping of benzimidazole-resistant and dicarboximide- resistant mutations in Botrytis cinerea using real-time polymerase chain reaction assaysPhytopathology98397404
9. R. Ben-yaacov, S. Knoller, G. A. Caldwell, J. M. Becker, Y. Koltin, 1994Candida albicans gene encoding resistance to benomyl and methotrexate is a multidrug resistance gene.Antimicrobial Agents and Chemotherapy. 38648652
10. M. D. Bolton, K. Birla, V. Rivera-varas, K. D. Rudolph, and G. A. Secor, 2012Characterization of CbCyp51 from field isolates of Cercospora beticola.Phytopathology102298305
11. K. J Brent, . 1995. Fungicide resistance in crop pathogens: How can it be managed? Global Crop Protection Federation, Brussels, Belgium. Available on-line at: http://www.frac.info/frac/publication/anhang/FRAC_Mono1_2007_100dpi.pdf. Last accessed 9/4/2012.
12. K. J. Brent, D. W. Holleman, 1998Fungicide resistance: the assessment of risk.FRAC Monograph II. Global Crop Protection Federation, Brussels, Belgium.

13. G. P. Canas-gutiérrez, M. J. Angarita-velásquez, J. M. Restrepo-flórez, P. Rodríguez, C. X. Moreno, and R. Arango, 2009Analysis of the CYP51 gene and encoded protein in propiconazole-resistant isolates of Mycosphaerella fijiensis. Pest Manag Sci 65892899

14. K. S. Chapman, G. W. Sundin, and J. L. Beckerman, 2011Identification of resistance to multiple fungicides in field populations ofVenturia inaequalisPlant Dis. 95921926

15. H. J. Cools, C. Bayon, S. Atkins, J. A. Lucas, and B. A. Fraaije, 2012Over-expression of the sterol 14α-demethylase gene (MgCYP51) in Mycosphaerella graminicola isolates confers a novel azole fungicide sensitivity phenotype. Pest Manag Sci 6810341040

16. H. J. Cools, B. A. Fraaije, S. H. Kim, and J. A. Lucas, 2006Impact of changes in the target 450CYP51 enzyme associated with altered triazole-sensitivity in fungal pathogens of cereal crops. Biochem Soc Trans 34: 1219-1222.

17. H. J. Cools, B. A. Fraaije, T. P. Bean, J. Antoniw, and J. A. Lucas, 2007Transcriptome profiling of the response of Mycosphaerella graminicola isolates to an azole fungicide using cDNA microarraysMol Plant Pathol 8639651

18. K. D. Cox, K. Quello, R. J. Deford, and J. L. Beckerman, 2009A rapid method to quantify fungicide sensitivity in the brown rot pathogen Monilinia fructicolaPlant Dis. 93328331

19. K. D. Cox, P. K. Bryson, and G. Schnabel, 2007Instability of propiconazole resistance and fitness in Monilinia fructicola. Phytopathology97448453

20. M. G. Cunha, D. M. Rizzo, 2003Development of fungicide cross resistance in Helminthosporium solani population from California. Plant Dis. 87798803

21. Del SorboG., Schoenbeck, H.-J. and De Waard, M.A. 2000Fungal transporters involved in efflux of natural toxic compounds and fungicidesFungal Genet. Biol. 30115

22. C. Delye, F. Laigret, and M. F. Corio-costet, 1997A mutation in the 14α-demethylase gene of Uncinula necator that correlates with resistance to a sterol biosynthesis inhibitor. Appl Environ. Microbiol. 6329662970

23. C. Delye, L. Bousset, and M. F. Corio-costet, 1998PCR cloning and detection of point mutations in the eburicol 14α-demethylase (CYP51) gene from Erysiphe graminis f sp. hordei, a 'recalcitrant' fungus. Curr Genet 34399403

24. A. R. Detweiler, J. M. Vargas, and T. K. Danneberger, 1983Resistance of Sclerotinia homoeocarpa to iprodione and benomylPlant Dis. 6762730

25. M. De Waard, A. C. Andrade, K. Hayashi, H. Schoonbeek, I. Stergiopoulos, and L. Zwiers, 2006Impact of fungal drug transporters on fungicide sensitivity, multidrug resistance and virulence. Pest Manag. Sci. 62195207

26. S. A. Dunbar, 2006Applications of Luminex xMAP technology for rapid, high-throughput multiplexed nucleic acid detection. Clin. Chim. Acta 3637182

27. P. S. Dyer, J. Hansen, A. Delaney, and J. A. Lucas, 2000Genetic control of resistance to the sterol 14α-demethylase inhibitor fungicide prochloraz in the cereal eyespot pathogen Tapesia yallundae. Appl. Environ. Microbiol. 6645994604

28. F. Faretra, and S. Pollastro, 1993Isolation, characterization and genetic analysis of laboratory mutants of Botryotinia fuckeliana resistant to the phenylpyrrole fungicide. Mycol. Res. 97620624

29. B. A. Fraaije, H. J. Cools, S-H. Kim, J. Motteram, W. S. Clark, and J. A. Lucas, 2007A. Novel, I. Substitution, V in the sterol 14α-demethylase (CYP51) of Mycosphaerella graminicola is differentially selected by azole fungicides. Mol Plant Pathol 8245254

30. B. A. Fraaije, J. A. Butters, J. M. Coehlo, D. R. Jones, and D. W. Hollemon, 2002Following the dynamics of strobilurin resistance in Blumeria graminis f. sp. tritici using quantitative allele-specific real-time PCR measurements with the fluorescent dye SYBR GreenI. Plant Pathol. 514554

31. A. Fontaine, F. Remuson, L. Fraissinet-techet, A. Micoud, R. Marmeisse, and D. Melayah, 2009Monitoring of Venturia inaequalis harboring the QoI resistance G143A mutation in French orchards as revealed by PCR assays. Pest Manag. Sci. 657481

32. FRAC, 2006. Mutations associated with QoI resistance. Available on line at http://frac.info/frac/work/Mutations%20associated%20with%20QoI%20resistance.pdf. Accessed 9/4/2012.

33. Frac, 2009. FRAC Code List: Fungicides sorted by mode of action (including FRAC Code numbering) http://www.frac.info/frac/publication/anhang/FRAC_CODE_LIST.pdf

34. M. Fujimura, N. Chiai, A. Ichiishi, R. Usami, K. Horikoshi, and I. Yamaguchi, 2000Sensitivity to phenylpyrrole fungicides and abnormal glycerol accumulation in Os and Cut mutant strains of Neurospora crassaJ. Pestic. Sci., 253136

35. J. L. Genet, G. Jaworska, and F. Deparis, 2006Effect of dose rate and mixtures of fungicides on selection for QoI resistance in populations of Plasmopara viticola.Pest Manag Sci 62188194

36. M. Grindle, 1982Fungicide-resistance of os mutants of Neurospora crassa. Neurospora Newslett., 291617

37. A Grube, ., D Donaldson, ., T Kiely, ., and L Wu, . 2011. Pesticide industry sales and usage. 2007 market estimates. 34 pp. Available at:http://www.epa.gov/opp00001/pestsales/07pestsales/market_estimates2007.pdf Last accessed 3/12/2012.

38. H. Hamamoto, K. Hasegawa, R. Nakaune, Y. J. Lee, Y. Makizumi, K. Akutsu, and T. Hibi, 2000Tandem repeat of a transcriptional enhancer upstream of the sterol 14 alpha-demethylase gene (CYP51A1) in Penicillium digitatum. Appl. Environ. Microbiol. 6634213426

39. H. Hayashi, M. A. Schoonbeek, and M. DeWaard, 2002Expression of the ABC transporter BcatrD from Botrytis cinerea reduces sensitivity to sterol demethylation inhibitor fungicidesPestic. Biochem. Physiol. 73110121

40. C F. Higgins, 1992ABC transporters: from microorganisms to man.Annu Rev. Cell. Biol. 867113

41. U. W. Hilber, and M. Hilber-bodmer, 1998Genetic basis and monitoring of resistance of Botryotinia fuckeliana to anilinopyrimidinesPlant Dis. 82496500

42. S. Ikuta, K. Takagi, R. B. Wallace, and K. Itakura, 1987Dissociation kinetics of 19 base paired oligonucleotide-DNA duplexes containing different single mismatched base pairs.Nucleic Acids Res. 15797811

43. H. Ishii, K. Yano, H. Date, A. Furuta, Y. Sagehashi, T. Yamaguchi, T. Sugiyama, K. Nishimura, and W. Hasama, 2007Molecular characterization and diagnosis of QoI resistance in cucumber and eggplant fungal pathogens.Phytopathology9714581466
44. H. Ishii, J. Fountaine, W-H. Chung, M. Kansako, K. Nishimura, K. Takahashi, and M. Oshima, 2009Characterisation of QoI-resistant and field isolates of Botrytis cinerea from citrus and strawberry. Pest Manag Sci 65916922
45. Y. Ito, H. Muraguchi, Y. Seshime, S. Oita, and S. O. Yanagi, 2004Flutolanil and carboxin resistance in Coprinus cinereus conferred by a mutation in the cytochrome b560 subunit of succinate dehydrogenase complex (Complex II) Mol. Genet. Genomics 272328335
46. L. M. Kawchuk, L. J. Hutchison, C. A. Verhaeghe, D. R. Lynch, P. S. Bains, and J. D. Holley, 2002Isolation of the β-tubulin gene and characterization of thiabendazole resistance in Gibberella pulicaris. Can. J. Plant Pathol. 24233238
47. J. E. King, and M. J. Griffin, 1985Survey of benomyl resistance in Pseudocercosporella herpotrichoides on winter wheat and barley in England and Wales in 1983Plant Pathology3427283
48. H. Koenraadt, and A. L. Jones, 1992aThe use of allele-specific oligonucleotide probes to characterization resistance to benomyl in field strains of Venturia inaequalis. Phytopathology 8213541358
49. H. Koenraadt, S. C. Somerville, and A. L. Jones, 1992bCharacterization of mutations in the beta-tubulin gene of benomyl-resistant field strains of Venturia inaequalis and other plant pathogenic fungiPhytopathology 8213481354
50. W. Köller, W. F. Wilcox, J. Barnard, A. L. Jones, and P. G. Braun, 1997Detection and quantification of resistance of Venturia inaequalis populations to sterol demethylation inhibitorsPhytopathology87184190
51. W. Köller, and W. F. Wilcox, 2001Evidence for the predisposition of fungicide-resistant phenotypes of Venturia inaequalis to a preferential selection for resistance to other fungicides. Phytopathology 91776781

52. W. Köller, D. M. Parker, and K. L. Reynolds, 1991Baseline sensitivities of Venturia inaequalis to sterol demethylation inhibitors. Plant Dis. 75726728
53. M. Kretschmer, M. Leroch, A. Mosbach, A-S. Walker, S. Fillinger, D. Mernke, et al2009Fungicide-driven evolution and molecular basis of multidrug resistance in field populations of the grey mould fungus Botrytis cinereaPLoS Pathogens 5113
54. P. J. Kuhn, 1984Mode of action of carboximides. Symp. Ser. Br. Mycol. Soc., 9155183
55. R. X. Latin, 2011A practical guide to turfgrass fungicidesThe American Phytopathological Society, 270 pp.
56. P. Leroux, M. Gredt, F. Remuson, A. Micoud, A-S. Walker, 2013Fungicide resistance status in French populations of the wheat eyespot fungi Oculimacula acuformis and Oculimacula yallundae. Pest. Manag. Sci.691526
57. P. Leroux, and A-S. Walker, 2011Multiple mechanisms account for resistance to sterol 14α-demethylation inhibitors in field isolates of Mycosphaerella graminicola. Pest. Manag. Sci. 674459
58. P. Leroux, M. Gredt, M. Leroch, A. S. Walker, 2010Exploring mechanisms of resistance to respiratory inhibitors in field strains of Botrytis cinerea, the causal agent of gray mold. App. Environ. Microbiol. 7666156630
59. P. Leroux, C. Albertini, A. Gautier, M. Gredt, and A. S. Walker, 2007Mutations in the cyp51 gene correlated with changes in sensitivity to sterol 14α-demethylation inhibitors in field isolates of Mycosphaerella graminicola. Pest. Manag. Sci. 63688699
60. P. Leroux, F. Chapeland, D. Desbrosses, and M. Gredt, 1999Patterns of cross-resistance to fungicides in Botryotinia fuckeliana (Botrytis cinerea) isolates from French vineyardsCrop Protect 18687697
61. P. Leroux, C. Lanen, and R. Fritz, 1992Similarities in the antifungal activities of fenpiclonil, iprodione and tolclofos-methyl against Botrytis cinerea and Fusarium nivale. Pestic. Sci. 36325329
62. P. Leroux, and M. Gredt, 1989Negative cross-resistance of benzimidazole-resistant strains of Botrytis cinerea, Fusarium nivale and Pseudocercosporella herpotrichoides to various pesticidesNetherlands J. Plant Pathol. 95121127

63. S. S. Lesemann, S. Schimpke, F. Dunemann, and H. B. Deising, 2007Mitochondrial heteroplasmy for the cytochrome b gene controls the levels of strobilurin resistance in the apple powdery mildew fungus Podosphaera leucotricha (Ell. & Ev.) E.S. Salmon. J Plant Dis Protect 113259266
64. K. E. Lesniak, T. J. Proffer, J. L. Beckerman, and G. W. Sundin, 2011Occurrence of QoI resistance and detection of the G143A mutation in Michigan populations of Venturia inaequalisPlant Dis. 95927934
65. C. Luo, X. Cox, K. D. Amiri, A. Schnabel, G. Occurrence, and detection of the DMI resistance-associated genetic element 'Mona' in Monilinia fructicola. Plant Dis. 9210991103
66. Luminex website: http://wwwluminexcorp.com/ TechnologiesScience/xMAPTechnology/ accessed 10/22/2012
67. H. Lyr, 1977Effects of fungicides on energy production and intermediary metabolism, 301332In M. Siegel and H. D. Sisler (ed.), Antifungal compounds. 2interactions in biological and ecological systems. Dekker M, New York, NY.
68. Z. Ma, and T. Michailides, 2005Advances in understanding molecular mechanisms of fungicide resistance and molecular detection of resistant genotype in phytopathogenic fungi. Crop Prot. 24853863
69. Z. Ma, T. J. Proffer, J. L. Jacobs, and G. W. Sundin, 2006Overexpression of the 14 alpha-demethylase target gene (CYP51) mediates fungicide resistance in Blumeriella jaapii. Appl. Environ. Microbiol. 7225812585
70. M. Mann, and O. N. Jensen, 2003Proteomic analysis of post-translational modifications.Nat. Biotechnol. 21255261
71. D. Mernke, S. Dahm, A. Walker, S. Lale, A. Ve, S. Fillinger, M. Leroch, and M. Hahn, 2011Two promoter rearrangements in a drug efflux transporter gene are responsible for the appearance and spread of multi- drug resistance phenotype MDR2 in Botrytis cinerea isolates in French and German vineyards. Phytopathology 10111761183
72. M. L. Metzker, 2010Sequencing technologies-the next generation. Nature Rev.Gen. 113146

73. M. Miguez, C. Reeve, P. Wood, and D. Hollomon, 2003Alternative oxidase reduces the sensitivity of Mycosphaerella graminicola to QoI fungicides.Pest. Manag. Sci. 6037

74. R. Nakaune, K. Adachi, O. Nawata, M. Tomiyama, K. Akutsu, and T. Hibi, 1998A novel ATP-binding cassette transporter involved in multidrug resistance in the phytopathogenic fungus Penicillium digitatum.Appl Environ Microbiol 6439833988

75. J. Nurmi, A. Ylikoski, T. Soukka, M. Karp, and T. Lövgren, 2000A new label technology for the detection of specific polymerase chain reaction products in a closed tubeNucleic Acids Res 15:28:e28

76. N. Ochiai, M. Fujimura, T. Motoyama, A. Ichiishi, R. Usami, K. Horijoshi, I. Yamaguchi, 2002Characterization of mutations in the two-component histidine kinase gene that confer fludioxonil resistance and somotic sensitivity in the os-1 mutants of Neurospora crassa. Pest Manag. Sci. 57437442

77. A. B. Orth, M. Rzhetskaya, E. J. Pell, M. Tien, 1995A. Serine, threonine) protein kinase confers fungicide resistance in the phytopathogenic fungus Ustilago maydis. Appl. Environ. Microbiol. 6123412345

78. A. B. Orth, A. Sfarra, E. J. Pell, and M. Tien, 1994Characterization and genetic analysis of laboratory mutants of Ustilago maydis resistant to dicarboximide and aromatic hydrocarbon fungicidesPhytopathology8412101214

79. M. Oshima, S. Banno, K. Okada, T. Takeuchi, M. Kimura, A. Ichiishi, I. Yamaguchi, M. Fujimura, 2006Survey of mutations of a histidine kinase gene BcOS1 in dicarboximide-resistant field isolates of Botrytis cinerea J. Gen. Plant Path. 726573

80. M. Oshima, M. Fujimura, S. Banno, C. Hashimoto, T. Motoyama, A. Ichiishi, and I. Yamaguchi, 2002A point mutation in the two component histidine kinase BcOS-1 gene confers dicarboximide resistance in field isolates of Botrytis cinerea. Phytopathology927580

81. S. S. Pao, I. T. Paulsen, and M. H. Saier, Jr. 1998Major facilitator superfamily. Microbiol. Mol. Biol. Rev. 62134

82. J. S. Pasche, L. M. Piche, N. C. Gudmestad, 2005Effect of the F129L mutation in Alternaria solani on fungicides affecting mitochondrial respirationPlant Dis. 89269278

83. J. S. Patel, S. W. Meinhardt, H. Sierotzki, G. Stammler, N. C. Gudmestad, and T. B. Adhikari, 2011A. Two-step, molecular detection method for Pyrenophora tritici-repentis isolates insensitive to QoI fungicides. Plant Dis. 9515581564
84. J. F. Petrosino, S. Highlander, R. A. Luna, R. A. Gibbs, and J. Versalovic, 2009Metagenomic pyrosequencing and microbial identification.Clin. Chem. 55856866
85. E. H. Pommer, and G. Lorenz, 1982Resistance of Botryis cinerea to dicarboximide fungicides-A literature Review. Crop. Prot. 1221230
86. K. L. Quello, K. S. Chapman, and J. L. Beckerman, 2010In situ detection of benzimidazole resistance in field isolates of Venturia inaequalis in IndianaPlant Dis. 94744750
87. M. A. Ramesh, R. D. Laidlaw, F. Durrenberger, A. B. Orth, and J. W. Kronstad, 2001The cAMP signal transudction pathway mediates resistance to dicarboximide and aromatic hydrocarbon fungicides in Ustilago maydis. Fungal Genet. Biol. 32183193
88. S. N. Rampersad, 2011A rapid colorimetric microtiter bioassay to evaluate fungicide sensitivity among Verticillium dahliae isolatesPlant Dis. 95248255
89. D. F. Ritchie, 1983Mycelia growth, peach fruit-rotting, capability, and sporulation of strains of Monilinia fructicola resistant to dichloran, iprodione, procymidone, and vinclozolin. Phytopath. 734447
90. J. Sambrook, E. F. Fritsch, T. Maniatis, 1989Molecular cloning: A laboratory manualCold Spring Harbor Laboratory Press, Cold Spring Harbor, NY), 2nd ed.
91. D. Sanglard, K. Kuchler, F. Ischer, J. L. Pagani, M. Monod, and J. Bille, 1995Mechanisms of resistance to azole antifungal agents in Candida albicans isolates from AIDS patients involve specific multidrug transporters.Antimicrob. Agents Chemother. 3923782386
92. D. Sanglard, F. Ischer, D. Calabrese, P. A. Majcherczyk, and J. Bille, 1999The ATP binding cassette transporter gene CgCDR1 from Candida glabrata is involved in the resistance of clinical isolates to azole antifungal agentsAntimicrob. Agents Chemother. 4327532765

93. G. Schnabel, and A. L. Jones, 2001The 14α- demethylase (CYP51A1) gene is overexpressed in Venturia inaequalis strains resistant to myclobutanil. Phytopathology 91102110
94. W. T. Schroeder, and R. Providenti, 1969Resistance to benomyl in powdery mildew of cucurbits. Plant Dis. Rep. 53271275
95. H. Sisler, 1988Fungicide action and fungicide resistance mechanisms. In Fungicide Resistance in North America. Delp, C.J. Editor 68
96. G. Stammler, J. Benzinger, Speakman. 2007A rapid and reliable method for monitoring the sensitivity of Sclerotinia sclerotiorum to boscalid J. Phytopathol., 155746748
97. G. Stammler, and J. Speakman, 2006Microtiter method to test the sensitivity of Botrytis cinerea to boscalid J. Phytopathol., 154508510
98. C. Stehmann, and M. A. De Waard, 1996Sensitivity of populations of Botrytis cinerea to triazoles, benomyl and vinclozolinEur. J. Plant Pathol. 102171180
99. H. Sierotzki, R. Frey, and J. Wullschleger, 2007Cytochrome b gene sequence and structure of Pyrenophora teres and P. tritici-repentis and implications for QoI resistancePest Manag Sci 63225233
100. I. Stergiopoulos, J. G. M. Van Nistelrooy, G. H. J. Kema, and M. A. De Waard, 2003Multiple mechanisms account for variation in base-line sensitivity to azole fungicides in field isolates of Mycosphaerella graminicola.Pest Manag Sci 5913331343
101. F. Suzuki, J. Yamaguchi, A. Koba, T. Nakajima, and M. Arai, 2010Changes in fungicide resistance frequency and population structure of Pyricularia oryzae after discontinuance of MBI-D fungicidesPlant Dis. 94329334
102. J. A. Tomlinson, I. Barker, and N. Boonham, 2007Faster, Simpler, More-Specific Methods for Improved Molecular Detection of Phytophthora ramorum in the Field. Appl. Environ. Microbiol. 7340404047
103. T Troy, . 2011. Fast-Growing Fungicide Markets. Farm Chemicals International.Sept.2011.http://www.farmchemicalsinternational.com/magazine/?storyid=3238

104. van den Bosch FPavely N, Shaw M, Hobbelen P and Oliver R. (2011The dose rate debate: does the risk of fungicide resistance increase or decrease with dose? Plant Pathol 60597606

105. S. M. Villani, and K. D. Cox, 2011Characterizing fenbuconazole and propiconazole sensitivity and prevalence of 'Mona' in isolates of Monilinia fructicola from New York. Plant Dis. 95828834

106. P. Vincelli, and N. Tisserat, 2008Nucleic acid-based pathogen detection in applied plant pathology. Plant Dis. 9266069

107. F Whitford, ., D Pike, ., F Burroughs, ., G Hanger, ., B Johnson, ., and D Brassard, . 2006. The Pesticide Marketplace: Discovering and developing new products. PPP-71. Available on-line at: http://www.ppp.purdue.edu/Pubs/ppp-71.pdf.

108. D Wilson, . and T Wicks, .. 2011. FungicideResistant Powdery Mildew In Australian Vines. Available on line, http://www.cropcare.com.au/Assets/564/1/ADVivando-2011FungicideResistantPowderyMildewInAustralianVines.pdf Accessed on 9/4/2012.

109. N. P. Wiederhold, J. L. Grabinski, G. Garcia-effron, D. S. Perlin, and S. A. Lee, 2008Pyrosequencing to detect mutations in FKS1 that confer reduced echinocandin susceptibility in Candida albicans. Antimicrob. Agents Chemother. 5241454148

110. H. M. M. J. Wood, J. A. Dickinson, P. S. Lucas, Dyer. 2001Cloning of the CYP51 gene from the eyespot pathogen Tapesia yallundae indicates that resistance to the DMI fungicide prochloraz is not related to sequence changes in the gene encoding the target site enzymeFEMS Microbiol. Lett., 196183187

111. R. A. Wyand, and J. K. M. Brown, 2005Sequence variation in the CYP51 gene of Blumeria graminis associated with resistance to sterol demethylase inhibiting fungicides. Fungal Genet. Biol. 42726735

112. O. Yarden, and T. Katan, 1993Mutation leading to substitutions at amino acids 198 and 200 of beta-tubulin that correlated with benomyl-resistance phenotypes of field strains of Botrytis cinerea. Phytopathology 8314781483

113. Y. Zhang, R. Lamm, C. Pillonel, S. Lam, and J. R. Xu, 2002Osmoregulation and fungicide resistance: the Neurospora crassa os-2 gene encodes a HOG1 mitogen-activated protein kinase homologue. Appl. Environ. Microbiol. 68532538

114. D. Zheng, G. Olaya, and W. Koeller, 2000Characterization of a laboratory mutant of Venturia inaequalis resistant to the strobilurin- related fungicide Kresoxim-methyl. Curr. Genet. 38148155

115. F. Zhu, P. K. Bryson, and G. Schnabel, 2012Influence of storage approaches on instability of propiconazole resistance in Monilinia fructicola. Pest Manag. Sci. 6810031009

116. L. H. Zwiers, I. Stergiopoulos, J. G. M. Van Nistelrooy, M. A. De Waard, 2002ABC transporters and azole susceptibility on laboratory strains of the wheat pathogen Mycosphaerella graminicola. Antimicrob. Agents Chemother.46 39003906

Chapter 2

Natural Products from Plants and Fungi as Fungicides

Marina D. Soković, Jasmina M, Glamočlija, and Ana D. Ćirić

University of Belgrade, Mycological Laboratory, Department of Plant Physiology, Institute for Biological Research, Belgrade, Serbia

INTRODUCTION

In early fifties of the twentieth century the agrochemical industry provided agriculture with a vast array of chemicals for crop protection, including fungicides. Random synthesis, biological screening and empirical optimization yielded many effective compounds (Cremlyn, 1991). Whereas it is generally acknowledged that the use of pesticides has large benefits to farmers, the present use of pesticides in agriculture also causes negative environmental (and health-related) effects to society. For example, during and after application of pesticides a substantial amount of it could end up in soil, ground- and surface water or air. These negative effects demands for an effective policy. Such policies have been initiated, both at the level of the individual Member States of the European Union and at the level of the European Union itself (Oppenheimer & Donnelly, 1997). Their selectivity between target organisms and plants is mainly based on differences in uptake. The more recently developed protective chemicals are more potent in terms of dose required to control the pest or disease, and in distinguishing between target and non-target organisms. They usually have a specific mode of action. Since selective compounds are specific site inhibitors in the metabolism of target organisms, the risk of developing resistance is high. This has occurred for a number of fungal plant pathogens (Delp, 1988). Although pathogenic microorganisms are mainly controlled chemically, the use of synthetic compounds is limited due to several undesirable aspects, which include carcinogenicity, teratogenicity, acute toxicity and the requirement of an extended degradation period with consequent development of environmental pollution problems. The new awareness of modern consumers about these problems has created a "green" consumer profile that demands the absence of synthetic chemicals in food production and preservation together with extended shelf life of the majority of food products. Fungal infections remain a therapeutic problem in many fields despite the availability of a number of treatments. Such diseases in humans have markedly increased during the past ten years, especially in immunocompromised patients. Consequently, up to 10% of hospital acquired systemic infections are caused by fungi. Altogether this forces the scientific community, agro–industry and pharmaceutical companies to search for natural compounds that will satisfy consumer requirements (Harvey, 2008). Furthermore, there is growing concern about chemicals for protection because of their undesirable side effects in humans, other target organisms and their behavior and fate in the environment (Jespers, 1994).

The total number of all known natural products is around one million, including both bioactive and inactive compounds, plants metabolites 600000, fungal metabolites 8600, microbial metabolites recognized until now is around 50000. It is an obvious question, where is the border in the diversity of natural products? The general needs of the human society are continuously increasing. We need every new compounds which may be useful for the human society. More food, new drugs, and other goods are highly necessary for the benefit of humankind. The only question is the existence of sufficient natural and technical resources to fulfill these demands. Fortunately, in the area of the research of bioactive microbial products it seems that the ever expanding scientific and technical possibilities are increasing together with the continuously widening needs of the human therapy, veterinary and agriculture. The problem really is not whether we would be able to discover further new useful microbial compounds, but rather how can we optimize and quickly and effectively apply the chances derived from the new discoveries. How can we pick up and use effectively the proverbial needle found in the haystack (Berdy, 2005). However, screening of more than a million substances in the last decade has resulted in the introduction of only a very limited number of compounds with novel modes of action and resistance. This explains the renewed interest of the chemical industry in natural compounds with a variety of unique characteristics, waiting to be exploited. Natural products derived from plants and fungi have traditionally been used in ethnomedicine. Throughout the development of both Western and Eastern civilizations, whole plants, fungi, their parts, derived compounds and extracts have functioned as sources of food and medicine, symbolic articles in religious and social ceremonies, and remedies to modify behavior. Plant and fungal extracts and compounds containing physiologically active biochemicals have immense potential for producing new agents of great benefit to mankind. In this context, systematic screening of secondary metabolites of folk herbs and fungi may result in the discovery of novel and effective antimicrobial compounds (Hussain et al., 2011). Recently, interest has been growing in natural products due to their availability, fewer side effects and less toxicity as well as better biodegradability when compared to other available antimicrobial agents and preservatives. Thus, plants and mushroom may offer great potential and hope. Consequently, natural products are attracting the attention of scientists because they are cheaper, safer, eco-friendly and within the reach of the current medical community. This paper gives an overview on the activity of plant and fungi derived extracts as well as their constituents against a wide variety of microfungi, methodology and potential uses (Figure 1.).

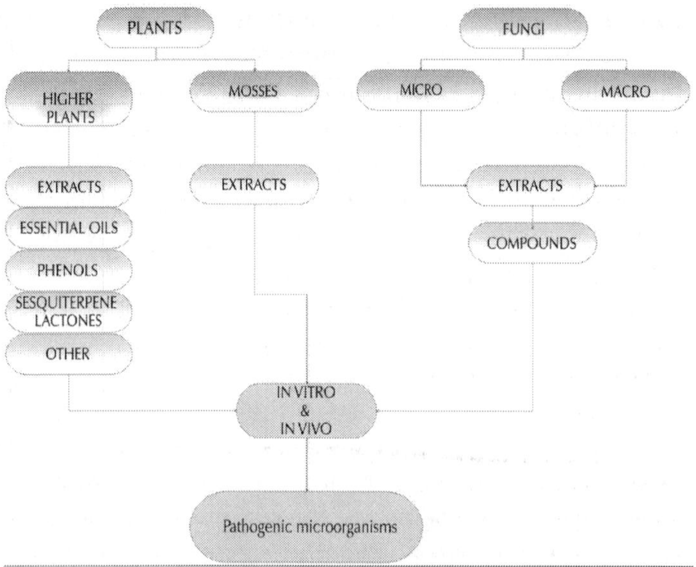

Figure 1: Our approach for testing of antifungal activity of natural products from fungi and plants.

METHODOLOGY

In order to test antifungal activities of natural products derived from plants and fungi few conventional and non-conventional methods were applied. Method is selected depending on the characteristics of extracts and compounds tested. A various number of plant, animal and human fungal pathogens were used. Growth cultures are conducted under optimal physical conditions for individual species. The growth of fungi was assessed visually or instrumentally. Two replicates were done for each compounds and the experiment was repeated two times.

Agar diffusion method is suitable for testing of antifungal activity of hydrophilic compounds which easily could be dispread trough to the agar medium. The compounds investigated were mixed with 0.01% Tween 20 surfactant and dissolved in molten MA medium. The fungal species were cultured for 7 days on Malt agar medium. Micromycetes were inoculated in the centre of Petri dishes and incubated for 21 days at 25° C. Mycelial growth was observed every 7 days and compared with the control. The commercial fungicide was used as a positive control (Ishii, 1995). The minimum inhibitory concentration (MIC) of compounds was determined when it achieved a complete stop in the growth of mycelium.

Microdilution method is suitable for testing small quantities of extracts, fractions or components, simultaneously in many different concentrations. In order to investigate the antifungal activity of the compounds, a modified microdilution technique was used (Hanel & Raether, 1988; Daouk et al., 1995). The fungal spores were washed from the surface of agar plates and adjusted with sterile saline to a concentration of approximately 1.0 x 10^5 in a final volume of 100 µl per well. Minimum inhibitory concentration (MIC) determinations were performed by a serial dilution technique using 96-well microtiter plates. The compounds investigated were dissolved in 5% DMSO solution containing 0.1% Tween 80 (v/v) (1 mg/ml) and added in broth Malt medium with inoculum. The microplates were incubated at Rotary shaker (160 rpm) for 72 h at 28 °C. The lowest concentrations without visible growth (at the binocular microscope) were defined as MICs. The fungicidal concentrations (MFCs) were determined by serial subcultivation of a 2 µl of tested compounds dissolved in medium and inoculated for 72 h, into microtiter plates containing 100 µl of broth per well and further incubation 72 h at 28 °C. The lowest concentration with no visible growth was defined as MFC indicating 99.5% killing of the original inoculum. DMSO was used as a negative control, commercial fungicides were used as positive controls.

Microatmosphere test, a slightly modified agar disc diffusion method, is perfectly suitable for the estimation of essential oil activity in vapor phase (Zollo et al., 1998). The assay was performed using mushroom pathogenic fungi. Petri dishes were filled with malt agar (MA), and then seeded with a 7 day-old mycelial culture of the tested fungi. The Petri dishes were then inverted and the determined amount of essential oils impregnated on sterile filter paper discs (4 mm) attached to the inverted lid (1 disc per lid). The Petri dishes were wrapped with parafilm along the rim, inverted and incubated for 21 days at 25 °C in an incubator. The results are presented as the diameter of the microorganism growth inhibition zone, or as the essential oil minimal inhibitory quantity (MIQ), which inhibits the total growth of microorganism. Commercial fungicides were tested as a control.

Bioautography is widely used for the examination of extracts. When the solvent evaporates, the broth and microorganisms are applied on the chromatographer paper or plates, and after the incubation time, growth is scrutinized. No growth is observed on the active spot components. Simultaneously, the components of the extract are eluted and identified. Different volumes of the mycelium extracts and pure compounds were dissolved in appropriate solutions. Ten micro liters of each sample were applied on TLC plates and sprayed either with freshly prepared fungal suspensions in nutrient broth (TSB). The plates were incubated for 18 h at 37°C and then sprayed with aqueous sol. 3% of p-iodonitrotetrazolium violet and stored for another 3 h. After this period plates were sprayed with 70% EtOH to stop fungal growth and were incubated for 36 h at 27 °C. White inhibition zones on a pinkish background were indicative of antimicrobial activity of tested extracts or compounds. The widths of these zones (mm) are the measure of efficiency and presented as minimal inhibitory concentration (MIC) (Pacher et al., 2001). As positive controls commercial fungicides have been used.

ANTIFUNGAL ACTIVITY AGAINST PLANT PATHOGENS

Fungal Extracts and Metabolites

The fungi constitute a very large group of organisms that are found everywhere and are of great importance to life on earth and to human society. This is mainly due to the many interactions among fungi and other organisms. Most fungi produce a wide variety of secondary metabolites with biological activity. A wealth of literature exist on the discovery and potential use of mycotics in agriculture (Berdy, 1980; Guterson, 1990). However, their exploitation in agriculture remained limited, because widespread application in crops might simultaneously select for resistance to these mycotics in human pathogens (Isono, 1990). The concept that substances derived from one living organism may affect another organsm is old. Some of the secondary metabolites that occur in fungi are fairly widespread, but many are confined to a few species. Hence, screening of further fungi species usually leads to the discovery of new bioactive secondary metabolites. The broad diversity of the fungi, as well as their easy acquisition makes them especially interesting for natural products screening program. Among fungal species, the various microscopic (filamentous) fungi (ascomycetes, fungi imperfecti, etc.) are the most frequent producers with about 6400 produced compounds. From the most common ascomycetes, namely from Aspergillus, Penicillium and Fusarium species 950, 900 and 350 compounds have been isolated, respectively. Besides them several other filamentous and endophytic species (Trichoderma, Phoma, Alternaria, Acremonium and Stachybotrys), are also good producers, each produces several hundreds of bioactive compounds. From higher fungal species – basidiomycetes or mushrooms – exemplified by Ganoderma, Lactarius or Agaricus species, altogether about 2000 active compounds have been derived. From yeasts, only 140 and from Myxomycetes (slime moulds) species 60 bioactive metabolites have been isolated. The chemically relatively simple fungal compounds, over the antibiotic activities frequently exhibit diverse biological effects, mainly phytotoxic and pharmacological activities. We should not forget, the great practical and historical importance of beta-lactams (penicillins, cephalosporins), the cyclosporin, and various statins (mevinolin, compactin, lovastatin, pravastatin, atrovastatin), which are all fungus derived compounds. Recently it is unquestionable that the interest to all types of fungal species, but mainly to endophytic and the so called marine fungi as possible sources of new bioactive compounds is highly increasing. The expansion of the very quick new screening methods led to the appearance of the increasing number of "unidentified" fungus as bioactive metabolite producers (over 250 new metabolites in the last two years), especially in the patent literature. It indicates the high speed of isolation, and identification/patenting process of new fungal products, and the long time need for taxonomical identification of new fungal species (Berdy, 2005). Particularly desirable is the discovery of novel prototype antimicrobial agents representing new chemical classes that operate by differet modes of action from existing agents and, consequently, lack cross-resistance to chemicals currently used. Kurobane et al., (1981) reported that Penicillium brefeldianum produces fulvic acid which possesses antiviral, antifungal, antioxidant and antibiotic activities. Maskey et al., (2003) isolated two active substances, 8-O-methylaverufin and 1,8-O-dimethylaverantin as new antifungal agents from Penicillium chrysogenum. Nam et al., (2000) found that 8-O-methylsclererotiorinamine isolated from Penicillium multicolor showed antimicrobial activity. Funiculosin, found in Penicillium funiculosum possesses antibiotic activity (Ando et al., 1969), as well as substance SQ 30,957, a new antibiotic produced by P. funiculosum (Singh et al., 1986). We recently started an investigation on the antifungal activity of fungal extracts and metabolites from both micro- and macrofungi. The extracts of 17 microfungi (Alternaria alternata, Cladosporium cladosporioides, C. fulvum, Fusarium sporotrichioides, F. trincintum, Paecilomyces variotii,Penicillium ochrochloron, P. funiculosum, Phoma magdonaldii, Phomopsis helianthi, Stachybotrys chartarum, Trichoderma viride, and five dermatomycetes, Epidermophyton floccosum, Microsporum canis, Trichophyton mentagrophytes, T. rubrum and T. tonsurans were tested against the yeast Candida albicans using the bioautographic assay test on TLC plates (Rančić, 2004).

While herbs are rather commonly used in the Western hemisphere, medicinal use of mushrooms, which has a long tradition in Asian countries, has also slightly increased in Europe during the last few decades. Although there has been extensive research on properties of medicinal mushrooms, their true potential is yet to be revealed. A number of compounds possessing significant antimicrobial activity have been isolated from polypore fungi. They provide a rich variety of active secondary metabolites and polysaccharides. Medicinal mushrooms such as Agaricus brasiliensis, Coprinus comatus, Coriolus versicolor, Ganoderma lucidum, Lentinula edodes, Phellinus linteus, and many others have traditionally been used as health foods or supplements for the prevention and cure of a range of diseases, including atherosclerosis, cancer, chronic hepatitis, and diabetes. The preventive and therapeutic effects of these mushrooms and their components have been well documented in mouse and rat model systems and in cancer cell lines. This has led to a considerable amount of knowledge about the effects of mushroom extracts and of their modes of action. It is generally accepted that mushroom extracts contain a variety of components, such as polysaccharides (i.e. glucans), small proteins, lectins and polyphenols, each of which may have its own biological or medicinal effect. The most common immunomodulatory action of mushroom are attributed to β-(1→3)-(1→6)-glucans, which have been studied in some detail (Smiderle et al., 2010). Vaz et al.,

(2011) described and compared the chemical constituents (phenol compounds, macronutrients, sugars, fatty acids, tocopherols and ascorbic acid) of four wild edible mushrooms widely appreciated in gastronomy:Armillaria mellea, Calocybe gambosa, Clitocybe odora, Coprinus comatus. Polysaccharides have emerged as an important class of bioactive substances, and many medicinal and therapeutic properties are attributed to them (Alquini & Carbonero, 2004). Trametes versicolor, Laetiporus sulphureus andGanoderma lucidum are just some of the known mushrooms with this potential. This alone has made them suitable candidates for critically needed new antibiotics and antimycotics (Zjawiony, 2004).Laetiporus sulphureus is a wood-rotting basidiomycete, growing on several tree species and producing shelf-shaped fruit bodies with a bright yellow fleshy margin. This recognizable pigmentation along with the fruit body form is responsible for the trivial name under which this fungus is known, and that is sulfur shelf (Weber et al., 2003). Even though it is recognized as a source of active compounds, and is widely used as a food among human, reports on the antimicrobial activity of L. sulphureus extracts are scarce (Turkoglu et al., 2007; Zjawiony, 2004). The potential barrier to everyday use of medicinal mushrooms as therapy is the manner in which the mushroom is consumed. Most research on fungi as potential antimicrobial agents is based on ethanol and methanol extracts of the fungal fruit body (Barros et al., 2007; Turkoglu et al., 2007). Consumption of products on this basis is of no practical use. To test the antifungal activity of extracts and metabolites of macrofungi we chose the wood-rotting basidiomycete, L. sulphureus, also named chicken of the woods. It is known for its nutritional value. An aqueous extract obtained from L. sulphureus was investigated for antimicrobial properties using a microdilution assay in vitro against seven fungi (four Aspergillus, two Penicillium species andTrichoderma viride). This extract showed strong activity against the tested microorganisms in a dose dependent manner (Šiljegović et al., 2011a). The presence and growth of microfungi in food may cause spoilage and result in reduction in quality and quantity. The presence of toxigenic fungi in foods stored for long periods of time is a potential hazard to human and animal health. Consumption of tomato products has been associated with a lower risk of developing digestive tract and prostate cancer. Therefore, preservation of tomato paste seems to be of great importance, both for the food industry, and for human well-being. We have used a methanol extract of L. sulphureus as an in vivoinhibitor of Aspergillus flavus growth in tomato paste. The results indicated complete inhibition of A. flavus growth in tomato paste for 15 days. An inhibition rate of 99.83% was achieved with 0.15 mg/ml of extract. Complete fungicide activity (100%) and no spore survival in the tomato product was recorded using 0.25 mg/ml of L. sulphureus extract in tomato medium. Since L. sulphureus is widely consumed as an edible macrofungus, its use as a natural preservative in tomato products can be considered as safe (Stojković et al., 2011a).

Plant Extracts and Metabolites

With increasing acceptance of traditional plants as an alternative form of health care the search for active compounds in plants becomes very important. Medicinal and aromatic plants have been employed for many centuries and they are mentioned in folklore from ancient times. After the advent of antibiotics in the 1950s, the use of plant derivatives as antimicrobials become virtually nonexistent to be rediscovered, as well as other alternative forms of medical treatments in the late 1990s (Cowan, 1999). There are several approaches to choosing sources of natural products for the discovery of potential antifungal compounds. One of approach is to investigate whole extracts of potential antifungal plants. Other approaches are to obtain biological material, which has not previously been studied for fractionation and testing, or some other sources. One strategy is to use ethnobotanical and/or chemical ecology clues to select which plants to sample (Duke et al., 2000). Here at first we will discuss the antifungal activities of plants extracts and after that some secondary metabolites derived from plants.

Plant Extracts

There are many reports concerning the antifungal activity of plant extracts, but we will mention only a few. Ushiki et al., (1996) found that root extracts from twelve medicinal plants displayed antimicrobial activity against certain pathogens of soil-borne plant diseases. Among these plants, Geranium pratense(Bigroot geranium) strongly inhibited the growth of Streptomyces scabies which causes common scab of potato. It was shown that geranin, isolated from Bigroot geranium roots possessed a 1.25% higher antimicrobial effect than streptomycin (Ushiki et al., 1997). Previous studies indicated that certain crops and vegetables contain antimicrobial substances in their roots, and that these substances directly suppress growth of the pathogen and development of the disease (Clarke, 1966; Masaoka et al., 1993;Naqvi & Chauhan 1980; Yoshihara et al., 1988). In a program to screen extracts from medicinal plants for fungicidal activity, it was found that aqueous extracts of Reynoutria suchalinensis (Polygonaceae) showed favorable protecting control of powdery mildew (Herger et al., 1988).

An ethanol extract of Phlomis fruticosa (Jerusalem sage) (Labiatae) tested by diffusion method inhibited Aspergillus niger, Penicillium ochrochloron, Trichoderma viride, Fusarium tricinctum andPhomopsis helianthi. Moreover, this extract had fungicidal activity against Cladosporium cladosporioides and Aspergillus ochraceus at a very low concentration (10-20 g/ml), (Ristić et al., 2000). Further investigation showed that when this extract was hydrolyzed with HCL and -glucosidases, which remove sugars from flavonoids, it possessed greater antifungal activities than the original ethanol extract. Lower antifungal activity of the whole ethanol extract may be due to the presence of some aglycones, unstable flavonoid glycosides (Soković et al., 2000). It is generally known that flavonoid glycosides show lower activity against the microorganisms than aglycones (Raoha et al., 2000). Strong antifungal activity of a dealcoholized extract of leaves of Cassia tora was obtained against Aspergillus niger (Mukherje et al., 1996). Sato et al., (2000) analyzed 29 plants extracts against Arthrinium sacchari and Chaetonium funicola. The ethanol extracts of fifteen plants showed antifungal activity, but Acer nikoense (Nikko maple), Glycyrrhiza glabra and Thea sinensis(Tsa) were the most effective plants in very low amounts. Whole, fresh involucral bracts of cardoon,Cynara cardunculus L. (Compositae), were extracted with EtOH and an aqueous suspension of this extract was partitioned successively with $CHCl_3$, EtOAc and n-BuOH, leaving a residual water extract. Each extract was evaluated for antifungal properties. Antimicrobial activity was estimated using a microdilution technique against Aspergillus niger, A. ochraceus, A. flavus, Penicillium ochrochloron, P. funiculosum, Trichoderma viride, Fusarium tricinctum and Alternaria alternata. All cardoon extracts were found to possess antifungal activity comparable with standard mycotics (Kukić et al., 2008). Antifungal assays of branched centaury Centaurium pulchellum (Gentianaceae) extracts and secoiridoid glycosides isolated from this extracts have been studied as potent bioactive compounds against five fungal species. Methanol extracts from both aerial parts and roots exhibited excellent antifungal (0.1-2 mg/ml) activity. Pure secoiridoid glycosides isolated from these extracts demonstrated very strong antifungal (0.001-0.1 mg/ml) activity (Šiler et al., 2010). The antifungal activity of methanol extracts of three different Labiatae species (Catmint), Nepeta rtanjensis, N. sibirica and N. nervosa (grown in vitro) against eight fungal species, was evaluated. All tested extracts showed significant antifungal activity, with that from, N. rtanjensis being the strongest (Nestorović et al., 2010).

Genuine mosses constitute a large group of nonvascular higher plants, consisting of about 14 000 species. Generally, bryophytes are not damaged by microorganisms, insects, snails, slugs, and other small mammals. Up to date, over several hundred new compounds have been isolated from bryophytes and their structures elucidated (Veljić et al., 2009). In spite of a number of secondary metabolites identified from various mosses, the chemical profiles of most species are insufficiently known or even unknown. Secondary metabolites from mosses, identified so far, include terpenoids, flavonoids and bibenzyls, and also derivatives of fatty acids. Mosses rich in flavonoids has been found to possess strong antimicrobial activity (Veljić et al., 2009). An ethanol extract of bryophyte, Bryum argenteum(silver moss), showed antifungal activity against two fungi (A. niger and P. ochrochloron) (Sabovljević et al., 2006). Our investigations also demonstrated that methanol extracts of selected genuine mosses (Pleurozium schreberi, Palustriella commutata, Homalothecium philippeanum, Anomodon attenuatus, Rhytidium rugosum, Hylocomium splendens, Dicranum scoparium and Leucobryum glaucum) possess antimicrobial activity when tested by the microdilution method (Veljić et al., 2008). When the antifungal activity of methanol extracts of the mosses Fontinalis antipyretica var. antipyretica, Hypnum cupressiforme and Ctenidium molluscum were analyzed, that of the first species showed strongest activity against the following micromycetes: Trichoderma viride, Penicillium funiculosum, P. ochrochloron, Aspergillus fumigatus, A. flavus and A. niger (Veljić et al., 2009). The antifungal activity of extracts of three bryophyte species, two mosses (Atrichum undulatum, Physcomitrella patens) and a liverwort (Marchantia polymorpha ssp. ruderalis), grown under natural conditions and in axenic culture, was evaluated by the microdilution method against five fungal species. Each bryophyte extracts was active against all fungi tested. In general, extracts made from material grown in laboratory (in vitro) conditions express stronger antifungal activity than those made from material from under natural conditions. Some of the fungi tested reacted similarly to both extracts (Sabovljević et al., 2011). The antimicrobial activity of a dimethyl sulfoxide extract of the moss Rhodobryum ontariense was evaluated by microdilution method against Aspergillus versicolor, A. fumigatus, Penicillium funiculosum, P. ochrochloron and Trichoderma viride. The extract was active against all the fungi tested but to varying degrees. This finding implies that R. ontariense could be considered as a promising material for natural antifungal products (Pejin et al., 2012 in press).

Plant Secondary Metabolites

Plants produce a diverse array of secondary metabolites, many of which have antifungal activity. Some of these compounds exist in healthy plants in biologically active forms. Others, such as cyanogenic glycosides and glucosinolates, occur as inactive precursors and are activated in response to tissue damage or pathogen attack. This activation often involves plant enzymes, which are released as a result of breakdown in cell integrity. Compounds belonging to the latter category are still regarded as constitutive because they are immediately derived from preexisting constituents (Mansfield, 1983). A large number of plant compounds have been reported to have antifungal activity. Well known examples include; flavonoids (Ćirić et al., 2011; Karioti et al., 2011; Weidenbörner & Jha 1997), lactones (Djeddi et al., 2007; Janaćković et al., 2002; Skaltsa et al., 2000a, 2000b; Vajs et al., 1999, 2004), proteins (Giudici et al., 2000), sulfur compounds (Ilić et al., 2012), cyanogenic glycosides and glucosinolates (Osbourn, 1996) and essential oils (Daouk et al., 1995; Džamić et al., 2010; Garg & Siddiqui, 1992;Glamočlija et al., 2006a, 2006b, 2009; Marinković et al., 2002; Mishra & Dubey, 1994; Müller-Riedau et al., 1995; Rančić et al., 2005; Shimoni et al, 1993; Soković, 2001, 2002, 2008a, 2008b, 2009a,2009b, 2009c; Stojković et al., 2011b, 2011b; Thompson, 1989).

Essential Oils: Essential oils from aromatic and medicinal plants have been known since antiquity to possess biological activity and constitute one of the most investigated groups of secondary metabolites. With growing interest in their use in the pharmaceutical and agrochemical industries, systematic examination of oils for these properties has become increasingly important. Over the last hundred years antimicrobial properties of common spice oils have been demonstrated (Bullerman et al., 1977) and many studies have been made on antifungal activities of essential oils (Daouk et al., 1995; Garg & Siddiqui, 1992;Glamočlija, 2006b, 2009; Kalemba & Kunicka, 2003; Mishra & Dubey, 1994; Müller-Riedau et al., 1995; Shimoni et al, 1993; Soković, 2001; Thompson, 1989). Thus, Maruzzela & Balter (1959) found that, out of 119 spice oils tested, 100 essential oils possessed an antagonistic effect to at least one of twelve phytopathogenic fungi and 50 of these compounds showed wide spectrum activity against all fungi tested. The essential oil of Origanum majorana exerted considerable inhibitory powers againstAspergillus flavus, A. niger, A. ochraceus, A. parasitcus and Trichoderma viride (Deans & Svoboda, 1990). A comparative study of the antifungal activity of essential oils extracted from thyme, rosemary, eucalyptus and mugwort was carried out by a group of investigators against 39 mold strains. The essential oil of thyme was found to be the most effective (Conner & Beuchat, 1984). Essential oils of allspice and cloves totally inhibited Trichoderma viride, Alternaria alternata, Fusarium oxysporum, Mucor circinelloides, Rhizopus stolonifer, Cladosporium cladosporioides, Aspergillus versicolor andPenicillium citrinum in a concentration of 2% (Schmitz et al., 1993). Essential oils from other plants such as Wormwood Artemisia afra, Lavender tree, Heteropyxis natalensis and Sweet gale Myrica gale,were found to have strong inhibitory effects against a broad spectrum of fungal species. The essential oil of Origanum syriacum showed very strong antifungal activity against Penicillium, Aspergillus andFusarium species (Daouk et al., 1995 and references cited therein). The essential oil of Soldier's herbPiper angustifolium was very effective effect against Aspergillus niger and A. flavus (Trillini et al., 1996). The antifungal activities of the essential oils from Lemon mint Monarda citriodora and Tea treeMelaleuca alternifolia were evaluated in vitro on fifteen common post-harvested pathogens of a variety of crops. Both essential oils exhibited a high level of antifungal activity, by direct contact and in the vapor phase. Oil from Lemon mint was generally more active than that from Tea tree, particularly against rapidly growing fungal species (Bishop & Thornton, 1997). Baratta et al., (1998) examined essential oil from eight commercial plants (Cinnamomum zeylanicu (Cinnamon), Cananga odorata(Ylang ylang), Ocimum basilicum (Sweet basil), Citrus limon (Lemon), Cymbopogon citratus (Lemon grass), Baswellia thurifera (Boswellia), Majorana hortensis (Marjoram) and Rosmarinus officinalis(Rosemary) and showed that all the oils tested were able to inhibit the growth of the common spoilage fungus, A. niger, even at a concentration of 1 l/ml broth, with the exception of lemon and rosemary oils which exhibited inhibitory effects on higher concentrations. Essential oils extracted from different parts of some angiosperms (Cedarwood Cedrus deodara and Ajwain Trachysremum ammi) were analyzed for fungitoxicity against the mycelial growth of Aspergillus niger and Curvularia ovoides,two fungi found in Vigna mungo. Since the essential oils from both plants exhibited fungitoxic properties, it may be possible to use them to control various fungi and exploit them as fungicides (Singh & Tripathi, 1999). The antifungal activities of four essential oils from spice (sage, thyme, oregano and savory) were analyzed against Fusarium oxysporum, Macrophomina phaseoli, Botrytis cinerea, Rhizoctonia solani, Alternaria solani and Aspergillus parasiticus. Earlier results showed weak activity for sage, while thyme, oregano and savory were active against all moulds tested (Ozcan & Boyraz, 2000). Different essential oils have antifungal activity against a wide range of fungi. Thus, considering the importance of these oils, 75 different essential oils were tested against A. niger. All the oils possessed antifungal activity (Pawer & Tacker, 2006). We examined a variety of essential oils from several plant families (Compositae, Labiatae, Lauraceae, Apiaceae, Cupressaceae, Poaceae, Illiaceae, Myrtaceae, Verbenaceae). Essential oils of Wild marjoram Origanum onites, Thyme-leaved savory Satureja thymbra, Greek sage Salvia fruticosa and S. pomifera subsp. calycina plants growing wild in Greece and their components; carvacrol, camphor, and 1,8-cineole, were assayed for antifungal activity against thirteen fungal species. The oils inhibited all fungi investigated. The highest and broadest activity was shown by oils containing the carvacrol (Wild marjoram and Thyme-leaved savory, while Greek sage was the least effective (Soković et al., 2002). The antifungal activity of essential oils from three Micromeria species: M. dalmatica, M. albanica and M. thymifolia was investigated against seven fungal species. The oils from all three Micromeria: (M. dalmatica-0.2-0.4 µl/ml, M. thymifolia-0.4-2 µl/ml) and particularly M. albanica (0.2-0.4 µl/ml) showed strong antifungal effects against all fungi tested: Aspergillus niger, A. ochraceus, Penicillium ochrochloron,Cladosporium cladosporioides, Fusarium tricinctum,

Phomopsis helianthi and Trichoderma viride(Marinković et al., 2002). The essential oil of Foeniculum vulgare was tested for antifungal activity against nine different plant pathogenic fungi. This oil was most effective against C. cladopsorioides andP. helianthi (0.8-4.0 µl/ml), more effective than bifonazole (Mimica-Dukić et al., 2003). The antifungal activity of oils from eight Stachys species were tested against fungal pathogens (A. niger andPenicillium ochrochloron). The greatest activity was obtained for S. scardica due to the high content of sesquiterpene hydrocarbons (69.3%) in this oil (Skaltsa et al., 2003). Essential oil from Juniperus excelsa (Cupressaceae) was evaluated for antifungal activity against twelve micromycetes from the following genera (Alternaria, Aspergillus, Cladosporium, Fusarium, Penicillium, Phomopsis andTrichoderma). This essential oil showed moderate-high antifungal characteristics with MIC of 8.0–40.0 µl/ml and MFCs of 10.0–50.0 µl/ml. The essential oil was the most effective against the phytopathogenic species Phomopsis helianthi while Trichoderma viride was the most resistant species treated with this essential oil. This essential oil exhibited higher antifungal activity than the commercial fungicide bifonazole (Soković et al., 2004). The antifungal activity of essential oils from different plant species of Lauracaea family (Aniba rosaeodora Rosewood, Laurus nobilis Bay tree, Sassafras albidumSassafras and Cinnamomum zeylanicum Cinnamon) was investigated against seventeen micromycetes. In order to determine fungistatic and fungicidal concentrations (MIC and MFC) both diffusion and microdilution tests were employed. Essential oil from cinnamon was the most effective as an antifungal agent, followed by rosewood and sassafras oils. Laurel oil possessed the lowest antifungal activity (Simić et al., 2004). The antimicrobial activities of essential oils isolated from Sweet cicely Myrrhis odorata, St. John›s wort Hypericum perforatum and Helichrysum arenarium were determinated by microdilution test. The greatest antifungal activity was observed for sweet cicely oil, while H. arenarium showed the lowest antifungal potential. Minimal inhibitory and fungicidal concentrations were 0.5-60 µg/ml. The oil of sweet cicely showed had higher activity than commercial product and was very effective against all fungi tested (Rančić et al., 2005).

Essential oils and their components from three spices of Salvia (Labiatae) were tested for antifungal activity in our laboratory. The antifungal activity of the corolla, calyx and leaf oils of Salvia brachyodon were analyzed. Antifungal activities of the essential oils were determined by the microdilution method against Aspergillus niger. The oil from the calyx possessed the strongest antifungal activity probably due to synergistic activity of all the components present. This oil contains high concentrations of sesquiterpenes and diterpenes, which were not found in the other samples investigated (Soković et al., 2005). Another species of Salvia genus, Clary sage (Salvia sclarea) was also tested as an antifungal agent. A concentration of 25 μl/ml showed fungicidal activity againstAspergillus, Penicillium, Trichoderma viride and Fusarium species. For Mucor mucedo the MFC was 15 μl/ml. Fungistatic and fungicidal activities of the oil against Cladosporium cladosporioides was recorded at concentrations of 2.5 μl/ml and 5 μl/ml. The most sensitive micromycetes wereCladosporium fulvum, Alternaria alternata, Phomopsis helianthi, and Phoma macdonaldii, where a concentration of 2.5 μl/ml was lethal (Džamić et al., 2008). The essential oil of Sardinian sage S. desoleana and its main components were investigated for antifungal activity against eleven micromycetes by macro- and micro-dilution methods. The essential oil and components investigated were diluted in ethanol and Tween 80 in both methods. We analyzed whole essential oil from Sardinian sage and its main components; linalyl acetate, 1,8-cineole and linalool, which together represent 52.71% of the total oil. Whole essential oil of Sardinian sage and linalool possessed strong antifungal activity, while 1,8-cineole exhibited only moderate potential. Linalyl acetate had the lowest antifungal potential. Values for MIC were lower with microdilution method (Soković et al., 2009a). The antifungal activity of Nepeta rtanjensis essential oil on mycelia growth has been determined by the diffusion method. It acted most efficiently against Alternaria species (0.6 μl/ml). Bipolaris spiciferaand Cladosporium cladosporoioides had MIC values of 1.0 μl/ml, whereas Trichoderma viride with an MIC value of 1.6 μl/ml against this essential oil (Ljaljević Grbić et al., 2007). Antimicrobial activity from Citronella Cymbopogon winterianus (Poaceae) and Karawya Carum carvi (Apiaceae) essential oils was investigated against seventeen fungal species. Oil of Karawya at concentration of 0.25–2.5 μl/ml stopped the growth of all tested micromycetes except for T. viride. For this resistant fungus, the concentration of Karawya oil had to be increased to 10 μl/ml. The essential oil from Citronella was less effective but still showed stronger antifungal activity than the commercial drug bifonazole. The MIC and MFC values of this oil were 1–20 μl/ml (Simić et al., 2008).

Essential oils isolated from the aerial parts during the flowering and vegetative phases, roots and seeds of the plant Portenschlagiella ramosissima were tested for antimicrobial activity against micromycetes (Aspergillus flavus, A. niger, A. versicolor, Penicillium ochrachloron, P. funiculosum andTrichoderma viride). All of the oils tested showed activity. The most effective was that isolated from the aerial parts of the plant during flowering, followed by the oil from the vegetative phase, seeds and roots. The most resistant fungal species were Aspergillus flavus and A. versicolor (Soković et al., 2008b). Essential oils of Star anise Illicium verum (Illiaceae) and Clavos Eugenia caryophyllata(Myrtaceae) were investigated as a potential antifungal agents. Star Anise oil exhibited fungicidal characteristics with MIC and MFC values of 2.5–25 μl/ml. Clavos oil showed strong antifungal activity at 0.1–2.5 μl/ml. The most resistant fungi were Trichoderma viride, Penicillium and Aspergillusspecies. The antimicrobial activity of star anise is mainly due to anethole while eugenol is responsible for antifungal effect of cloves oil. The authors raised the possibility that interactive effects of other compounds present in smaller quantities may also contribute. Thuse, both oils, but especially clove, showed powerful antifungal activity (Džamić et al., 2009). The antifungal activity from essential oils of wild carrot Daucus carota L. (Apiaceae) collected in Serbia were tested. The antifungal activity of oils from the ripe fruits, unripe fruits, flowers, root, leaves, and stem, were examined against eight fungal strains (Fusarium sporotrichoides, Fulvia fulvum, Trichoderma viride, Penicillium ochrochloron, P. funiculosum, Aspergillus ochraceus, A. flavus and A.fumigatus) by the microdilution technique. The essential oil of unripe fruits manifested the strongest antifungal potential followed by oils from ripe fruits, roots, stems, leaves and flowers. These oils were more efficient than the commercial drug bifonazole and much more active than ketoconazole. The most prominent biological activity was exhibited by the essential oils from ripe and unripe fruits of wild carrot oil (Soković et al., 2009c). The chemical composition and effectiveness of the essential oil obtained from Echinophora spinosa(Apiaceae) was tested on different fungi. The most resistant fungal species were Penicillium ochrochloron and P. funiculosum while Trichoderma viride was the most sensitive. This essential oil tested showed higher antifungal potency against T. viride than the commercial drugs bifonazole and ketoconazole (Glamočlija et al., 2011a). The essential oil of Lippia alba Bushy lippia (Verbenaceae) is reported as an antifungal agent against human pathogenic microorganisms but few articles concern its use for green mould control. We determined the antifungal activity of Bushy lippia essential oil against green molds (Aspergillus ochraceus, A. niger, A. versicolor. A. fumigatus, Penicillium ochrochloron, P. funiculosum and Trichoderma viride) as an alternative to synthetic fungicides. Microdilution assays evaluated the essential oil MIC and MFC. Bushy lippia essential oil has MIC of 0.3-1.25 mg/ml and MFC of 0.6-1.25 mg/ml. Bushy lippia essential oil is classificated as citral type and the results indicate that it is a potential alternative to synthetic fungicides (Glamočlija et al., 2011b). The chemical composition and antimicrobial activities of the essential oils isolated from Pink savory Satureja thymbraand Black thyme Thymbra spicata (Labiatae) were compared. The oil of Black thyme possessed higher antifungal potential than Pink savory S. thymbra oil. A. versicolor and A. fumigatus were the most sensitive species, while P. ochrochloron was most resistant to these oils. Both oils showed much greater antifungal activity than a commercial antifungal agent (Marković et al., 2011). The essential oil of Seseli montanum subsp. tommasinii was tested for antifungal activity on four fungal species (Aspergillus ochraceus, A. fumigatus, Penicillium ochrochloron and Trichoderma viride). It showed moderate activity against all the tested fungi, but activity against A. fumigatus and T. viride was stronger than that of bifonazole. In the case of A. fumigatus, which is a very common and invasive pathogen, this is important due to the rising problem of fungal resistance to antifungal agents (Šiljegović et al., 2011b). We have studied the antimicrobial activity of Seseli species, examining essential oils obtained from the aerial parts of S. anuum (Milosavljević et al., 2007), S. globiferum fruits (Stojković et al., 2008a), S. globiferum aerial parts (Janaćković et al., 2011) and flowers of S. rigidum(Stojković et al., 2009). Differences in their activity were found. The essential oil from aerial parts ofS. anuum showed activity against twelve fungi in the range of 12.5 to 50 µl/ml. That from fruits of S. globiferum had the strongest antifungal activity with the MICs and MFCs in the range of 0.5-50 µl/ml. Oil from the aerial part of S. globiferum showed significant activity against micromycetes (2.5-10 µl/ml). The essential oil from the aerial parts of S. montanum subsp. tommasinii was more active than that from the aerial parts of S. anuum. In the case of P. ochrochloron and T. viride this activity was twice that of S. anuum oil. On the other hand, the essential oil from S. rigidum had greater antifungal activity than that of S. montanum subsp. tommasinii against A. fumigatus and P. ochrochloron(Šiljegović et al., 2011b). The results of our investigation of the antifungal activities of essential oils from sixteen aromatic and medical plants and their components (Bitter orange Citrus aurantium,Lemon C. limon, Hyssopus officinalis, Lavender Lavandula angustifolia, Wild lavender L. stoechas,Chamomile Matricaria chamomilla, Melissa Melissa officinalis, Peppermint Mentha piperita, Spearmint M. spicata, Sweet basil, Rusmary Rosmarinus officinalis, Sage Salvia officinalis, Sardinian sage S. desoleana, Clary S. sclarea, Thyme Thymus vulgaris and T. tosevii) by the microdilution method showed that Thymus essential oils were the most effective in tests in vitro, while those fromCitrus species and S. officinalis showed the lowest antifungal activities (Soković, 2001, 2009b). The same essential oils plus Oregano Origanum vulgare oil were assayed for inhibitory activity against major pathogens of the button mushroom, Agaricus bisporus, i.e. the fungi Verticillium fungicola andTrichoderma harzianum. The highest and broadest activity was shown by oregano and thyme oils with very low active concentrations (0.05-5.0 μg/ml) (Soković & van Griensven, 2006a). All the essential oils mentioned previously were also tested by microatmospheric method in vitro against 5 different isolates of Mycogone perniciosa (causal agent of wet bubble disease) from Agaricus bisporus. Essential oils which contained phenol components (thymol and carvacrol), (oregano, Greek oregano O. heracleoticum, Pink savory S. thymbra, thyme and T. tosevii with MIC values of 0.001 to 0.7 μl/disc and MFC of 0.1 to 1.0 μl/disc) showed significantly stronger antifungal potential than those with high alcohol contents (Spearmint and Peppermint with MIC 1.0 μl/disc and MFC 2.5 μl/disc), and those oils with high keton contents (S. pomifera with MIC 1.0 μl/disc and MFC 15.0 μl/disc and H. officinaliswith MIC 5.0 μl/disc and MFC 25.0 μl/disc), and especially than oils with monoterpenic hydrocarbons as dominant components (Lemon with MIC 4.0 μl/disc and MFC 5.0 μl/disc Bitter orange with MIC 0.7 μl/disc and MFC 1.0 μl/disc and Lavender MIC 1.5 μl/disc and MFC 2.5 μl/disc). The fungicide prochloraz showed much lower antifungal potential than all the oils tested, with MIC 5.0 μl/disc and MFC 50.0 μl/disc (Glamočlija, 2006a, 2006b, 2009). Also, there is only limited information in the literature on the antifungal activity of essential oils in vivo. The experiments in vivo are in relation with several problems of application of essential oils. Those related to the volatility of the oils and their poor solubility in water must be resolved before trials are performed in vivo. The persistence of a volatile oil on the treated plant and consequently its protection against pathogens is of short duration. The solubility problem means that organic diluents may have to be used, with the risk of environmental pollution and even phytoxicity. In an attempt to overcome these problems, some experimental formulations of essential oil and camphor, the most active components of Sage against Botrytis cinereawere prepared using as excipient a polymeric matrix obtained by the graft polymerization of acyclic monomers on gelled starch. Besides being absolutely harmless to plants, these polymers showed the capacity to form aqueous dispersions of the oil that were homogenous and adhered well to the surface of the leaves in some preliminary tests. Once the water had evaporated, the dispersions left a solid film which could be reduced to a hydrogel on the laminae of the leaves, from which the oil was gradually released in concentration related to the humidity of the microenvironment. This guarantees an extended duration of effective concentrations of the oil on treated plants (Moretti et al., 1998). The same researchers analyzed the effects of essential oil from Sage and camphor on tomato plants infected withBotrytis cinerea. A significant reduction of infection was found, but they not able to eliminate entirely the appearance of spots on the leaves. However, this encouraged further research in this field, with other essential oils which possessed greater antifungal activitiy in vitro. The effects of eleven plants essential oils for protecting maize kernel against Aspergillus flavus were studied. The optimal doses for maize protection, the influence of combinations of oils, residual effects and toxicity of the essential oils to maize plants were determinated. The principal constituents of eight essential oils were tested for their ability to protect maize kernels. Essential oils of Cinnamon, Peppermint, Sweet basil, Oregano, Telaxys ambrosioides, Syzygium aromaticum and thyme totality inhibited fungal development on maize kernel. Thymol and methoxycinnamaldexyde significantly reduced maize grain contamination. No phytotoxic effect on germination and corn growth was detected with any of these oils (Montes-Belmont & Carvajal, 1998). Reddy et al., (1998) showed in vivo that essential oils of thyme exhibited antifungal activity against Botrytis cinerea and Rhizopus stolonifer, two common pathogens of Fragaria anamassa.

Among a variety of oils tested in vitro against the pathogenic fungi Mycogone perniciosa, Oregano oil was singled out as the best. Although disease caused by M. perniciosa is routinely controlled with different fungicides, it remains a constant threat. In order to find alternative preventive methods we evaluated the antifungal activity of Oregano oil in vivo in a mushroom growing unit. To treat experimentally induced mushroom disease in the growing house we tested the antifungal activity of oregano oil when applied in casing soil. The most favorable results were achieved with 2% of oregano oil and simultaneous application of the spores suspension when oregano oil completely inhibited the growth of M. perniciosa. As essential oils are largely nontoxic and easily biodegradable we advise on disinfection of commercial casing soil with 2 % oregano oil before applying the casing to the compost (Glamočlija, 2006a, 2007). The chemical composition and antimicrobial activity of essential oils ofVitex agnus-castus L. and their main constituents in vitro and in vivo showed that these oils could be used equally in vitro and in vivo (Figure 2.). The oils from all plant parts possessed great antifungal potential against eight fungal plant pathogens. Using the same technique 1,8-cineole and α-pinene (dominant compounds) showed very high antifungal potency as well. As 1,8-cineole was the predominant constituent of the oils, we tested it in vivo. Randomly chosen apples were treated with 1,8-cineol solution and infected with Aspergillus niger in order to provoke Aspergillus rot in apples. As apple fruits were treated with an acceptable amount of 1,8-cineole, we suggest it as a potent agent for preventing apple rot caused by A. niger and its application as a bioactive compound to control A. nigerinfection during apple storage (Stojković et al., 2011b).

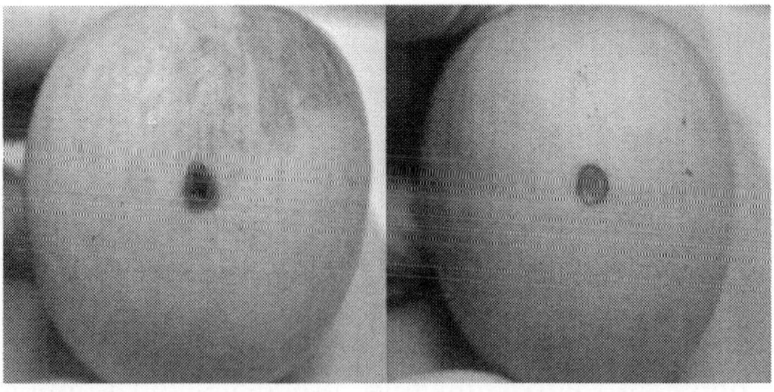

Figure 2: Treatment of Aspergillus rot in apples with 1,8-cineol.

Essential Oils Components: We will also discuss here the antifungal activities of essential oil components. Eugenol has exhibited very strong antifungal activity against Absida glauca, Aspergillus nidulans, A. niger, Colletotrichum capsici, Fusarium monoliformae, Pestolotia psidi and Rhizopus nadssus, while caryophyllene inhibitedA. glauca. Cineole showed favorable activity against Fusarium monoliformae and Alternaria alternataand good to moderate activity against the remaining fungi tested. The chillie crop in India suffers from rips tot and die Back diseases caused by C. capsici. The growth of this fungus has very successfully been inhibited by cumaldehyde and eugenol. Cineole, cumalaldehyde and eugenol also very satisfactorily inhibited growth of F. monoliformae, P. psidi and R. nadssus. F. monoliformae causes localized "Softrot" of apical tissue, head blight, scab and root rot of wheat and other cereals, beke disease of rice and twisted top of sugarcane. These compounds could be used for inhibition of A. alternata, which attacks crucifers such as mustard, cauliflower, knol-knol and radish. Cumalaldehyde may also find application as an inhibitory agent against the growth of P. expansum which causes infection if it gains a foothold in injured tissues. It also causes soft rot in apple fruit (Garg & Siddiqui, 1992). The essential oil of Foeniculum vulgare and components (anethole, fenhone and camphor) were screened for antifungal activity against nine plant pathogens. Anethole possessed the greatest activity (1.3-2.8 μl/ml), then fenchone (3.7-6.0 μl/ml), while camphor had the lowest antifungal effect (2.8-9.7 μl/ml) (Mimica-Dukić et al., 2003).

Essential oil components (β-caryophyllene, β-caryophyllene oxide, α-pinene, cadinene, linalool) from eight Stachys species were analyzed for antifungal activity. Linalool exhibited the greatest activity (0.03 mg/ml), while the lowest effect was observed for β-caryophyllene (0.3 mg/ml) (Skaltsa et al., 2003). We investigated the antifungal activity of limonene against fourteen fungal species using micro- and diffusion tests. Limonene showed antimicrobial activity against all fungi tested, in concentrations of 8.0-13.0 l/ml in the diffusion and 6.0-10.0 l/ml in the microdilution method. Differences in MICs obtained by these two methods could be attributed to low solubility of limonene in the agar medium used in the diffusion method. Limonene showed stronger antifungal potential than bifonazole, especially in the microdilution assay when the component was dissolved in Tween (Rančić et al., 2003). Phenol compounds were the most active among the components investigated. Limonene and linalyl acetate were less effective against micromycetes. All essential oils and components were investigated against twelve fungi: Aspergillus niger, A. ochraceus, A. versicolor, A. flavus, A. terreus, Alternaria alternata, Penicillium ochrochloron, P. funiculosum, Cladosporium cladosporioides, Trichoderma viride,Fusarium tricinctum and Phomopsis helianthi (Soković, 2001, 2009b). The components linalyl acetate, linalool, limonene, α-pinene, β-pinene, 1,8-cineole, camphor, carvacrol, thymol and menthol were assayed for inhibitory activity against three major pathogens of the button mushroom, Agaricus bisporus, the fungi Verticillium fungicola and Trichoderma harzianum. The highest and broadest activity was shown by carvacrol and thymol with very low MIC and MFC values (0.02-1.5 μl/ml), while linalyl acetate and limonene possessed the lowest activity (MIC/MFC 5.0-11.0 μl/ml) (Soković & van Griensven, 2006a). The essential oil of Critmum maritimum and its components (α-pinene and limonene) possessed antifungal activity against the mycopathogen M. perniciosa tested by the microatmospheric method. MIC for α-pinene was 5 μl/disc, and MFC 10 μl/disc, while limonene showed higher antifungal activity with MIC 1 μl/disc, and MFC 5 μl/disc (Glamočlija et al., 2009) (Table 1.).

It can be seen that growth of the tested fungi responded diversely to the essential oils and their components, which indicates that different components may have different modes of action or that the metabolism of some fungi is able to overcome the effect of the oil or adapt to it. Terepenic compounds inhibit electron transport, proton translocation, phosphorylation steps and other enzyme-dependent reactions or act on the cell membrane. Their antifungal activity will depend on the chemical composition of the cell wall and on the structure of the terpenoid molecules. Terpenic hydrocarbons are water insoluble and revealed poor activity, while among the water soluble compounds vanilin, piperonal and camphor, were not remarkably active, whereas the non-aromatic ester borneol acetate showed antiseptic effects. Aliphatic alcohols, such as linalool or citronellol and ketones like pipertone or carvone exhibited antifungal properties. Phenol compounds showed very strong antifungal activity in spite of their relative low capacity to dissolve in water (Knobloch et al., 1988). The most active terpenoids were found among phenols, followed by aldehydes and ketones, alcohols and hydrocarbons. Thymol and carvacrol were the most effective compounds which causing total inhibition of oxidative phosphorylation. The ability of terpenoids to inhibit the reactions described above arises both from lipophilic properties, which enables them to dissolve in the cytoplasmic membrane, and from their functional groups, which interfere with enzyme structure (Griffin et al., 1999; Knobloch et al., 1988;Shelef, 1983; Soković, 2001). Studies of antimicrobial activity of essential oils and their components showed that, terpene acetates and hydrocarbons tended to be relatively inactive, regardless of their structural type, and that this inactivity appears to be closely related to their limited hydrogen bonding capacity and water solubility. Ketones, aldehydes and alcohols showed activity but with differing specificity that was not always defined by the functional group present but was associated with hydrogen-bonding parameters in all cases (Griffin et al., 1999). Our results concerning the antifungal activity of many essential oils and their components indicate different efficacy. Also, the modes of action of essential oils differ among fungal species. The strong antifungal activity of some oils (Menthaspecies, Thyme, Oregano) can be explained by the high percentage by their high percentage of active components such as menthol, thymol, carvacrol. For the remaining oils, no significant correlation between antifungal activity and relative amounts of the major components has been found. This suggests that the components present in large proportions are not necessarily responsible for a great share of the total activity. Different antifungal activity exhibited by an oil, compared with those of it major components, can be explained by either synergistic effect of diverse components in the oil and/or by the presence of other components that may be active even in small concentrations (Soković & van Griensven, 2006a). To examine the problem of a lack of unified criteria in greater depth, we can look particularly at studies of the antimicrobial activity of essential oils. Janssen et al., (1987) reviewed the characteristics of complex mixtures as well as the techniques used to evaluate them and concluded that many results are difficult to compare as the test methods differed so widely. They proposed that in future the strain number of the tested microorganism, the composition of the essential oil and the conditions under which it was obtained be included as an integral part of the report.

Recently, Kalemba & Kunicka (2003) reviewed the classical methods commonly used to evaluate the antibacterial and antifungal activities of essential oils, including the agar diffusion method (paper disc and well), the dilution method (agar and liquid broth) and turbidimetric and impedimetric monitoring of microorganism growth in the presence of these oils. Besides drawing conclusions about factors that influence the antimicrobial activity of essential oils in vitro and their mechanisms of action, they included an overview of the susceptibility of human and foodborne fungi towards different essential oils and their constituents. The most relevant ones, which included oils from Thyme, Oregao, Mint, Cinnamon, Sage and Clove, have antimicrobial properties. Other criteria were the study of plants used as preservatives, as well as examination of the use of spices as antimicrobial agents. The cited criteria seem sufficient to justify the studies, but Ríos & Recio, (2005 and references cited therein) believe that research should be focused on achieving definitive knowledge about the plant and its properties. With increasing acceptance that the chemical diversity of natural products is well suited to provide core scaffolds for future antimicrobial agents, there will be more developments in the use of novel natural products and chemical libraries based on natural products (Harvey, 2008). The methodology employed is another point that needs to be considered in depth. For non-polar extracts, the use of diffusion techniques is probably inadequate, although many reports employing these techniques have been published. Solid dilution techniques are suitable for studying plant extracts or nonpolar compounds. Only when the amount of sample available is small is the use a diffusion techniques possibly more appropriate (Rios & Recio, 2005).

Our results showed that the MICs for essential oils and their components are generally higher, in disc-diffusion assays and with diffusion methods than with the microdilution method. Poor activity can be explained by low water solubility of the oil and its components, which limits diffusion through the agar medium in the disc diffusion and agar diffusion methods. Only more water-soluble compounds, such as 1,8-cineole, diffuse into the agar. The hydrocarbon components either remain on the surface of the medium or evaporate. This could be the reason for better results obtained using the microdilution method. Also, essential oil and their components showed greater antifungal activity when diluted in Tween 80. Both MICs and MFCs were lower in the microdilution than in the diffusion method, especially when Tween 80 was employed. Non-ionic emulsifiers, such as Tween 20 or 80, are relatively inactive when tested alone and have been widely reported as useful emulsifying agents (Soković, 2001; Soković et al.,

2002; Soković & van Griensven, 2006a). We observed that some oils and compounds acted not only as fungicidal agents but also inhibited sporulation of different fungi (Glamočlija et al., 2006a, 2006b; 2007; Soković & van Griensven, 2006a). Treatment was not only effective in solution or by contact, but even in a vapor treatment their were very effective enabling fungal growth to be inhibited by a smaller amount of essential oil while also acting as a potent inhibitor of sporulation. Vapor concentration and the duration of exposure are important. The gaseous contact activity was demonstrated primarily by the maximum vapor concentration at an early stage of incubation. Maintaining a high vapor concentration for long periods of time appeared to be unnecessary. Essential oil vapors might serve to control proliferation of moulds that are now treated with other sanitizing agents. Oils and their components have high vapor pressures and are relatively volatile. Solutions and emulsions used in the form of sprays with or without a carrier therefore represent the preferred form in which these agents should be applied to large areas of casing soil surface with minimal effort. Also evaporation by heating could be considered. An additional advantage of the volatile of essential oils is that no or only little residue will be left on the product after treatment. Our own experience leads us to propose the use of the microdilution method, carried out in microtiter trays, which involves a low workloads for a larger number of replicates and the use of small volumes of test substance and growth medium (Soković 2001, Soković et al., 2006a, 2009b). On other hand, for investigation of some plants and fungal extracts and separation of fractions and highly pure compounds bioautographic methods on TLC plates were recommended. This qualitative techniques will only give an idea of the presence or absence of substances with antimicrobial activity in very small amounts. The method was useful for screening plants and fungi for antimicrobial activity and for the bioassay-guided isolation of natural antimicrobial compounds. Bioautography allows easy localization of activity even a matrix as complex as that derived from natural products. Comparison chromatograms developed under identical conditions and visualized using suitable chromogen reagents can provide useful information about the nature of active compounds (Figure 3.) (Rančić et al., 2006, Ćirić, 2010).

Considerable changes in legislation have been made and there are increasing consumer trends for more natural alternatives to chemical fungicides (Brul & Coote, 1999). The use of essential oils is particularly advisable because herbs and spices are common plant additives. Among the natural antimicrobials that were tested in our laboratory the essential oils of Oregano and Thyme, as well as their components, carvacrol and thymol were the most promising. Addition of various plant derived antimicrobials in combination should improve both the spectrum of activity and the extent of inhibition due to synergistic effects. Thus, combination of these compounds might have even higher potential. The use of essential oils is limited, and possible reasons for this may be their strong smell and taste when used at effective doses (Skandamis & Nychas, 2000). Although the majority of essential oils are classified as Generally Recognized As Safe (GRAS) (Kabara, 1991), their use in foods as preservatives is often limited due to flavor considerations, as effective antimicrobial doses may exceed organoleptically acceptable levels. Therefore, there is an increasing demand for accurate knowledge of the minimum inhibitory (effective) concentrations (MIC) of essential oils to enable a balance between sensory acceptability and antimicrobial efficacy (Lambert et al., 2001).

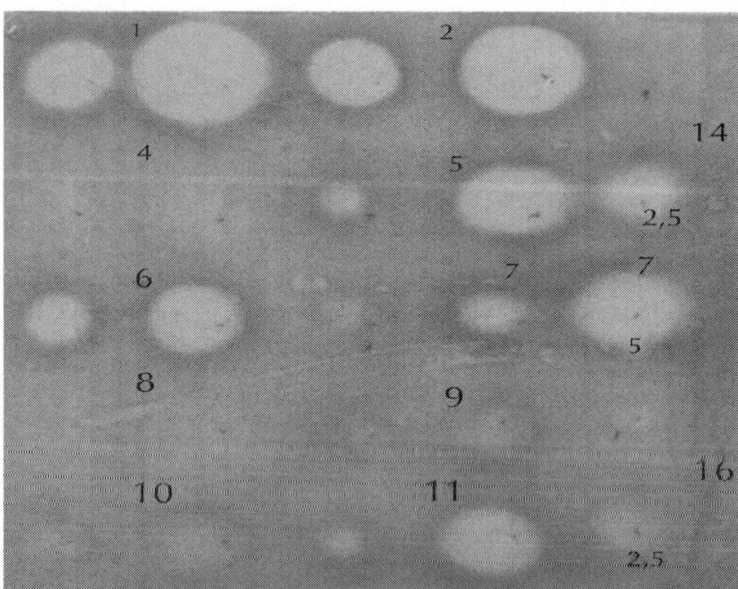

Figure 3: Antifungal activity of sesquiterpene lactones using a bioautographic method on TLC plate.

Table 1: An overview of antifungal activities of natural products from fungi and plants

Group of compounds	Species	Methods	Activity	**References**
Fungal extracts	Cladosporium fuivum Fusarium sporotrichioides Penicillium ochrochloron P. fitniculosum Phomopsis heliattthi Stachybotrys chartarum Trichoderma viride Microsporum canis Trichophytott ntbrum T. mentagrophytes T. tontonsurans	bioautographic assay on TLC plates bioautographic	50-100μg/ml 50-100 μg/ml 50-100 μg/ml 100 pg/ml 100 pg/ml 100 pg/ml 100 }.1g/ml 100 pg/m1 100 μg/ml 50-100 μg/ml	Raneie et al., 2006
	Laetiporus sulphureus	microdilution	0.30 mg/ml	Siljegovie et al., 2011a
plant extracts	Phlomis fruticosa	diffusion	10-20 μg /m1	Ristie et al., 2000
	Cynara cardunculus	microdilution	1.0-1.5 mg/ml	Kukie et al., 2008
	Centauriurn pulcluqlum	microdilution	0.1-2.0 mg/ml	iler et al., 2010
	Nepeta rtanjensis, N.sibirica, N. nervosa	microdilution	25-100 μg/ml	Nestorovia et al., 2010
mosses	Bryium argenteutn	microdilution	0.29-0.52 μg/ml	Sabovljevie et al., 2006

	Pleurozium schreberi Palustriella commutate Homalothecium philippeanum Anomodon attenuates Rhytidium rugosum Hylocomium splendens Dicranum scoparittm Leucobryutn glaucum	microdilution diffusion	0.5-10 mg/ml 1.0 mg/disc 0.5-10 mg/ml 1.0 mg/disc 0.5-10 mg/ml 0.5-10 mg/ml 0.5-10 mg/ml 0.5-10 mg/ml 1.0 mg/disc 0.5-10 mg/ml 0.5-10 mg/ml	Veljia et al., 2008
	Fontinalis antipyretica var Antipyretica Hypnum cupressifortne Ctenidium mollusc=	microdilution	2.5-5 mg/ml 5 mg/ml. 5 mg/ml	Veljie et al., 2009
	Atrichum undulatum, Physcomitrella patens, Marchantia polymorpha ssp. ruderalis	microdilution	0.1-2 mg/ml 0.1-2 mg/ ml 0.25-1 mg/ml	Sabovljevie et al., 2011
	Rhodobryum ontariense	microdilution	0.25-1 mg/ml	Pejin et al., 2012
essential oils	Origanum onites, Satureja thymbra, Salvia fruticosa, S. pomifera subsp. calycina	microdilution	0.05-25 µl/m1	SokoviC et al., 2002
	Micromeria dalmatica, M. albanica, M. thymifolia	diffusion	0.2-2.0 µl/ml	Marinkovie et al., 2002
	Foeniculum vulgare	microdilution	0.84.0 µl/ml	Mimica-Dukie et al, 2003
	Stachys sp.	microdilution	0.01-1.0 mg/ml	Skaltsa et al., 2003

	Aniba rosaeodora Laurus nobilis Sassafras albidum Ciiznamomti In zeylanicum	diffusion microdilution	0.5-7.5 µl/ml 1.0-20 µl/ml 10-40121/ml 10-50 µl/ml 5-15 µ1/ml 5-30 µl/ml 0.1-1 µl/ml 0.1-2.5 µl/ml	Simic et al, 2004
	Juniperus excelsa	microdilution	8-50 µl/ml	Sokovic a al., 2004
	Myrrhis odorata Hypericum perforatum Helichrysum arenarium	microdilution	0.5-2.5 µg/ml 15-30 µg/ml 10-60µg/ml	
	Salvia brachyodon	microdilution	15-40 µl/ml	
	Salvia sclarea	microdilution	2.5-25 µl/ml	
	Salvia desoleana	diffusion microdilution	1.5-6 µl/ml 1.5-6 µl/ml	
	Nepeta rtanjensis	diffusion	0.6-1.8µl/ml	
	Cymbopogon winterianus Carum carvi	microdilution	0.5-20 µl/ml 0.25-20 µl/ml	
	Portenschlagiella ramosissima	microdilution	50-200 µl/ml	
	Eugenia caryophyllata Illicium verum	microdilution	0.1-2.5 µl/ml 2.5-25 µl/ml	
	Daunts carota	microdilution	2-150 µl/ml	

Morphophysiological Changes in Fungi Due to Inhibition Activity by Essential Oils: Dematiceous fungi are characterized by the presence of the dark brown pigment – melanin within their cell wall structure.

Melanins are negatively charged, hydrophobic biopolymers of high molecular weights. They are typically brown or black and formed by the oxidative polymerization of phenol or indolic compounds in organisms from all biological kingdoms, including fungi. Fungal melanins are usually found in the cell walls of spores, sclerotia, mycelia or fruiting bodies. They enable fungi to survive adverse environmental conditions by protecting them from oxygen free radicals, UV radiation and wall-degrading enzymes produced by antagonist microbes (Butler et al., 2001).

Many human pathogenic fungi contain melanin within their cell wall structure (e.g. Aspergillus fumigatus, A. nidulans, A. niger, Alternaria alternata, Cladosporium carionii, Cryptococcus neoformans, Exophiala jeanselmei, Fonsecaea compacta, F. pedrosoi, Hendersonula toruloidii, Histoplasma capsulatum, Paracoccidioides brasiliensis, Penicillium marneffei, Phaeoannellomyces wernickii, Phialophora richardsiae, P. verrucosu, Sporothrix schenckii, Wangiella dermatitidis). For several of these fungi, melanin has been described as a virulence factor due to its ability to reduce a pathogen's susceptibility for killing by host antimicrobial mechanisms and by influencing the host immune response. Due to the protective role of fungal melanin, dematiceous fungi are extremely difficult to treat with antifungal drugs (Nosanchuk & Casadevall, 2006). Plant secondary metabolites could be a suitable alternative for the treatment of fungal infections in the light of increasing fungal resistance to commercial antifungal agents (Vivek et al., 2009). Essential oils are known to cause morphophysiological changes in fungi through a lack of sporulation, depigmentation and aberrant development of conidiophores (e.g. Sharma & Tripathi, 2008; Moreira et al., 2010). We tested several essential oils for this purpose. Using the diffusion method it was observed that essential oils ofHyssopus officinalis, Thyme, T. tosevii, Spearmint and Peppermint induced changes in some morphophysiological characteristics of the fungi Trichoderma viride, Penicillium ochrochloron andAspergillus niger. Depigmentation of the colonies of A. niger was noted. Untreated control colonies were black, blue, green and the mycelium was well developed, while the colonies treated with oils were white with sparse mycelium. In the treated A. niger, P. ochrochloron and T. viride cultures, mycelium is rare, the conidiogenic apparatus was atypical, the vesicles deformed, phialides were abnormal and other variations included lack of sporulation, visible loss of pigmentation and aberrant development of conidiophores (Soković, 2001). Nepeta rtanjensis an essential oil showed the ability to interfere with all stages in the reproduction cycle of the human pathogenic fungus Bipolaris spicifera: conidia germination, mycelial growth and intensity of sporulation which is demonstrated with radial mycelial growth inhibition. Thus, inhibition of conidia germination and low conidia production were demonstrated in the treated samples. The most significant documented orphophysiological changes inB. spicifera included demelanization (bleaching) and an aberrant conidial apparatus (Ljaljević Grbić et al., 2011).

Flavonoids: Flavonoids are another group of secondary metabolites with great antifungal potential. Besides other biological activities, have been shown to be active against microorganisms. At least in some cases their presence might serve as a chemical barrier to invading microorganisms. Since they are natural compounds and possess highly specific antimicrobial activity, flavonoids may be an alternative to conventional fungicides in the control of plant diseases caused by fungi. Twenty-five flavonoids were examined for their effect on the mycelial growth of a crop pathogen, Verticillim albo-atrum. The minimum inhibitory concentrations (MIC) for the two most active compounds, flavone and flavanone. Other flavonoids inhibited hyphal growth and some compounds were ineffective at the highest concentration used. Active compounds did not share a common pattern of substitutions. The unsubstituted flavonoids were stronger growth inhibitors and, in most cases, increasing the number of substittions (hydroxylation, methoxylation and glycosylation) resulted in the loss of antifungal activity (Picman et al., 1995). The fungicidal activity of two isoflavones, one isoflavanone and seven isoflavans was tested against Aspergillus repens, A. amstelodami, A. chevalieri, A. flavus and A. petrakii(Weidenbörner et al., 1989). While the isoflavones showed low activity, the two isoflavans were highly inhibitory (Weidenbörner & Jha, 1997). Two naturally occurring isoflavones, genistein and biochanin A, and their dihydroderivates (isoflavanones) as well as nine perhydrogenated isoflavones (isoflavans) were tested for their effects against Rhizoctonia solani and Sclerotium rolfsii (Weidenbörner et al., 1990). All the isoflavonoids of the biochanin A series showed high antifungal activity. Genistein isoflavan and the other isoflavans with two hydroxyl groups and one methoxy group were fungitoxic, while isoflavans with two or three methoxy groups were almost inactive

(Weidenbörner & Jha, 1997), although earlier results demonstrated that isoflavans, generally possess higher activity than the corresponding isoflavones and isoflavanones.

Since the individual unsubstituted flavonoids showed strong antifungal activity, various mixtures have been tested against fungi occurring on grain to enhance the fungicidal potential of each substances by anticipating synergy. In general a combination of flavones and flavanones in different proportions was most effective. However, it was interesting that a mixture containing flavonoid molecules with one methoxy group and several hydroxy groups in general exhibited higher activity than a mixture containing only hydroxylated flavonoids. It becomes obvious that combinations of several suitable flavonoids (depending on the number, kind and location of the substitutents) may result in even greater increase in antifungal potential. Consequently, lower active concentrations may make flavonoids more attractive as natural protectants (Silva et al., 1998; Weidenbörner & Jha, 1997). Concerning the antifungal activity of flavonoid glycosides, it should be noted that no substantial effect could be detected (Weidenbörner & Jha, 1997). Krauze-Baranowska et al., (1999) found that cupressuflavone and 4′-O-methylcupressuflavone, isolated from leaves of Cupressocyparis leylandrii, possessed antifungal activity against Alternaria alternata, Cladosporium oxysporum, Fusarium culmorum and F. avenaceum. Matshumoto & Tahara, (2001) separated ampelosin, a flavonol, from Salix sachalinensisleaves and reported antifungal activity against Cladosporium herbarum.

The antifungal activities of many phenol compounds isolated from Holm Oak Quercus ilex leaves, belonging to the classes of flavonoids, proanthocyanidins, and phenol acids, have been examined against fourteen fungal species (Karioti et al., 2011). Two coumarins, scopoletin and isoscopoletin, two simple phenol acids, protocatechuic acid and isovanillic acid and one flavonoid, eriodictyol separated from the aerial parts of Centaurea spruneri, showed fungistatic activity at 0.259–2.38 µmol/ml and fungicidal at 0.69–2.6 µmol/ml against all fungi tested. The flavonoid, eriodictyol, possessed the greatest antifungal activity in the range of 0.694–1.388 µmol/ml for MIC and MFC, while the activity of protocatechuic acid was lower (0.65–1.3 µmol/ml for MIC and 1.3–2.6 µmol/ml for MFC). The simple phenol acids, protocatechuic acid expressed the lowest antifungal activity (Ćirić et al., 2011). The inhibitory activity of flavonoids generally decreased with the increasing number of substitutions on the molecule, and the strongest inhibitors were unsubstituted compounds. The most active flavonoids, flavone and flavanone, have excellent potential as new natural antifungal agents (Picman et al., 1995).

Sesquiterpene Lactones: Sesquiterpene lactones are natural products present in many families of plants, but mostly distributed in the family Compositae. They display a wide spectrum of biological activity, one of the most important of which is antifungal activity. The general mechanism of action is considered to be alkylation of biological nucleophiles such as cysteine (cys) and glutathione or sulfhydryl-containing systems, phosphofructoki-nase and glycogen synthetase by a, b-unsaturated carbonyl structures in a Michael-type addition (Koukoulitsa et al., 2005). This group of compounds was analyzed for potential antifungal activity, in vitro against Aspergillus niger, A. ochraceus, Penicillium ochrochloron, Trichoderma viride, Fusarium tricinctum, Phomopsis helianthi and Cladosporium cladosporioides. Guaianolides from Centaurea nicolai were found to be highly active (0.8-25 g/ml) (Vajs et al., 1999). Sesquiterpene lactones isolated from Centaurea achaia, C. thessala and C. attica also exhibited excellent antifungal activity against Aspergillus niger, A. ochraceus, A. versicolor, A. flavus, Penicillium ochrochloron, P. funiculosum, Trichoderma viride, Fusarium tricinctum, Phomopsis helianthi, Alternaria alternata andCladosporium cladosporioides with germacranolides providing the greatest antifungal activity (Skaltsa et al., 2000a; 2000b).

The fungicidal activities of 36 natural and synthetic sesquiterpene lactones with guaianolide, trans-germacranolide, cis-germacranolide, medampolide, and eudesmanolide carbon skeletons were evaluated against the phytopathogenic fungi Colletotrichum acutatum, C. fragarie, C. gloeosporioides, Fusarium oxysporum, Botrytis cinerea and Phomopsis sp. by Wedge et al., (2000). Dehydrozaluzanin showed the highest antifungal activity due the presence of an ,-unsaturated carbonyl group in the cyclopentananone ring. In addition to the previously isolated sesquiterpene lactones, 11,13-dihydrocnicin and 11,13-dihydro-19-desoxycnicin, the aerial parts of Centaurea pullata afforded three minor sesquiterpene lactones, namely, a new germacranolide, 8R-O-(4-acetoxy-5-hydroxyangeloyl)-11,13-dihydrocnicin, and two new eudesmanolides, 8R-O-(4-hydroxy-2-methylenebutanoyloxy)-11,13-dihydrosonchucarpolide and 8R-O-(4-hydroxy-2-methylenebutanoyloxy)-11,13-dihydro-4-episonchucarpolide. The antimicrobial activity of all previously mentioned compounds and some newly isolated sesquiterpene lactones (a novel elemanolide with an α-methyl-γ-lactone moiety, 8α-O-(4-hydroxy-2-methylenebutanoyloxy)melitensine, in addition to other four known sesquiterpene lactones with the same ring, melitensine, 11β-dihydrosalonitenolide, 8α-hydroxy-11β-13-dixidroxy-4-epi-sonchucarpolideand8α-hydroxy-11β-13-dihydroxy-onopordaldehyde) from Centaurea pullata was tested against eight fungal species, using a microdilution method. All compounds evaluated showed greater antifungal activity than the positive controls used. Moreover, the pharmacokinetic profile of these compounds was investigated using computational methods and was in agreement with our in vitro data (Djeddi et al., 2007, 2008). Three linear sesquiterpene lactones, anthecotulide, hydroxyanthecotulide and acetoxyanthecotulide were isolated from the aerial parts of Anthemis auriculata together with five known flavonoids, taraxa-20(30)en-3â-ol and methyl vanillate. Comparing these results with previously published data (Konstantinopoulou et al., 2003) concerning the antimicrobial potential of sesquiterpene lactones of A. altissima, it was concluded that linear lactones are more active. This differentiation in antimicrobial activity could be explained in terms of solubility, as linear lactones are more lipophilic than the oxygenated eudesmanolides and germacranolides of A. altissima (Theodori et al., 2006). Nine sesquiterpene lactones, anthemin A, 1α-hydroxydeacetylirinol-4 α,5β-epoxide, anthemin C, tatridin A, 1-epi-tatridin B, anthemin B, 6-deacetyl-β-cyclopyrethrosin, elegalactone A and 1β,4 α,6trihydroxyeudesm-11-en-8 -12-olide were isolated from the aerial parts ofAnthemis melanolepis. All sesquiterpene lactones showed an inhibitory effect against almost all fungi tested (Saroglou et al., 2010), which adds to evidence from previous studies concerning related compounds isolated from other Anthemis species (Konstantinopoulou et al., 2003).

The biological activity of sesquiterpene lactones is generally attributed to the alkylating property of the -methylene--lactone moiety. Moreover, the presence of other alkylating sites (epoxides and conjugated carbonyl groups) may enhance their biological activities. Lipophilicity seems to play an important role in antifungal activity. Since the chemical composition of fungal cells walls is highly lipophilic, they generally provide a strong barriers against penetration of hydrophobic compounds and transport of polar compounds through the outer lipid layer. According to Skaltsa et al., (2000a) an inverse relationship exists between polarity and antifungal activity for sesquiterpene lactones in general. Their polarity decreases in the order eudesmanolides elemanolides germacranolides. Some of the differences between the responses of Verticillium albo-atrum to flavonoids and sesquiterpene lactones suggest these two groups of plant metabolites have different modes of action. Thus, inhibition of mycelial growth by sesquiterpene lactones remained relatively constant during incubation times of 24, 48 and 72h. This suggests that these lactones affected the metabolism of the pathogen, slowing its growth, but that the fungus evidently did not produce specific enzymes to degrade the lactones. In contrast, flavonoids and especially flavanone, significantly reduced hyphal growth during the first 24h by up to 100% but growth was much less inhibited during the next 48h of incubation which indicated that the flavonoids were subject to degradation by enzymes produced by the pathogen (Picman et al., 1995).

Other Compounds Tested: Inhibition of certain thiol-containing enzymes in microorganisms by the rapid reaction of thiosulfinates with thiol groups was assumed to be the main mechanism involved in the antimicrobial effect of allicin. The mode of action of allicin on the fungal cell has not yet been elucidated but it is assumed to act on thiol enzymes as in other microorganisms. Other requirements such as molecular accessibility and lipophilicity seem to play an important role for in their antifungal activity (Yamada & Azuma, 1997). Antifungal activities of allicin and related organo-sulfur products obtained by microwave-assisted transformation of allicin in ethanol were studied in our laboratory against eight fungi: Aspergillus flavus, A. niger, A. fumigatus, Penicillium funiculosum, P. ochrochloron, Trichoderma viride, Candida albicans and C. kruzei. The mixture of transformation products of allicin was analyzed using liquid chromatography-mass spectrometry (LC-MS) and consisted of ajoenes, vinyldithiins and diallyl disulfide. Allicin showed very powerful antifungal activity with minimum inhibitory concentration (MIC) at 0.001 to 0.008 mg/ml and MFC at 0.004 to 0.03 mg/ml. The transformation products of allicin also possessed very strong antifungal activity, but less than allicin (Ilić et al., 2012).

Synergistic Effect of Different Natural Compounds and Fungicides: Fungi that are pathogenic to plants which produce antimicrobial compounds often have greater tolerance to these natural compounds in vitro than do nonpathogens of these plants, suggesting that resistance may be a prerequisite for infection. Nevertheless, many other factors will be required for fungal pathogenicity in addition to resistance to host antimicrobial compounds. Although in vitro tests of antifungal activity usually involve individual purified compounds, phytopathogenic fungi may often be exposed to more than one antifungal compound simultaneously during infection of plants. The combination of different compounds may be synergistic. The combined therapy is used with the aim of expanding the antimicrobial spectrum, minimizing toxicity, preventing the emergence of resistant mutants during therapy, and obtaining synergistic antimicrobial activity. Some steroidal glycoalkaloids have low antifungal activity when tested separately, but exhibit pronounced synergistic activity when mixed in plants (Osbourn, 1996). Synergism between ketoconazole and Agastache rugosa oil againstBlastichizomyces capitatus was reported, as well as between Pelargonium graveolens oil and amphotericin B plus ketoconazole on strains of Aspergillus sp. (Silva & Fernades, 2010). Thus, studies on the interactions between natural products and antifungal drugs have also multiplied in recent years, indicating the importance of elucidating types of interactions, which can be favorable, such as in synergism, or harmful, as in antagonism. Our results showed that proanthocyanidins (procyanidin and prodelphinidin) isolated from Holm oak when combined with bifonazole and ketoconazole, increased the activity of both conventional fungicides (Karioti et al., 2011). Compounds tested showed higher activity than bifonazole and ketoconazole, against A. fumigatus and A. niger. Almost all compounds tested exhibited good ability to inhibit fungi, even much better than commercial antifungal agents used as reference drugs. Even more, compounds showed few times higher activity than bifonazole and especially than ketoconazole. Also, we tested the synergistic effect of mixture of several essential oils and found that some combinations could possess great antifungal potential. The following mixture of essential oils: Thyme, Peppermint, Sage, Hyssopus officinalis, Sweet basil and Lavander showed greater antifungal potential in vitro than each oil separately (Soković, 2001). Structurally unrelated compounds are also likely to have combined and possibly synergistic antifungal activity. Therefore, this study supports the potential use of a weak antifungal products together with another compound to increase its activity. This type of finding could further boost the use of medicinal plants, extracts or natural products, either alone, combined or together with mycotics.

ANTIFUNGAL ACTIVITY AGAINST ANIMAL AND HUMAN PATHOGENS

Among animal and human pathogens, dermatomycetes are the main cause of dermatomycoses (infections of the hair, skin, and nails), superficial infections that are not life threatening but are chronic and cause considerable morbidity. The unpleasant side effects of therapy including nausea, abdominal pain and itching, and its toxicity, can limits its therapeutic use in many cases (Shin & Lim, 2004).

Despite the advancements of science and technology, surprisingly the development of novel and efficient antifungal drugs is still lagging behind due to the very fact that fungi are also eukaryotic and have mechanisms similar to human beings. Hence it becomes very difficult to develop an antifungal agent that is more specific in targeting the fungi alone without any damage to human beings. For successful treatment of the disease, proper diagnosis of the disease is always essential. The treatment is chosen based on the infection site, etiological agent and penetration ability of the drug. The penetration ability and retention in the site of infection of the agent determines its efficacy and frequency of utility. Since the dermatomycetes reside in the stratum corneum especially within the keratinocytes, the antifungal agents should have a good penetrating ability (Lakshmipathy & Kannabiran, 2010). Therefore, the development of more effective and less toxic antifungal agents is required for the treatment of dermatomycosis. Essential oils play a great role in these investigations, the majority have good penetration possibilities the lipophilic properties of oil components might have also aided in the ability of the oil to penetrate the plasma membrane, strong antifungal activity and if they are used in active (MIC and MFC) concentration they are not harmful for animals and humans.

Previous studies in vitro and in vivo on investigations of the antifungal activity of essential oils from some medicinal and aromatic plants, Origanum vulgare subsp. hirtum, Spearmint, Lavander, andSalvia fruticosa, indicated that they could be employed as effective antifungal agents (Adam et al., 1998). Our selection of plants for evaluation in our study was based on traditional usage for treatment of infectious diseases (Sokmen, 1999). The in vivo evaluation of antifungal activity of several essential oils and their components was tested for the therapeutic potency against experimentally induced dermatomycoses in rats (2-month old male Wistar rats), using the most frequent dermatomycetes,Trichophyton mentagrophytes, T. rubrum and T. tonsurans. Essential oil of Lavander exhibited therapeutic activity after 13 days of treatment. The group of rats treated with Sweet basil oil were cured after 25 days of treatment. The shortest period of currency was observed at animals treated with Sage-12 days. The longest period of treatment was observed at rats treated with oils of Bitter orange and Lemon, 45 days. The main essential oil components were also used as a potential antifungal agents. Linalool showed antifungal activity after 32 days of treatment, while limonene needs 50 days for this activity. Rats treated with 1,8-cineole were cured after 40 days. Camphor exhibited therapeutic and antifungal activity after 14 days of treatment. The best antifungal activity was observed for menthol, which showed therapeutic potential after 10 days of treatment (Soković, 2001). We examined the antifungal activity of essential oil from Mentha x piperita and menthol. The oil completely cured the animals infected with T. mentagrophytes within 15 days, with T. rubrum within 30 days and with T. tonsurans within 29 days. Menthol possessed higher therapeutic and antifungal activities than the essential oil, as it cured the animals within 10 days. Also, menthol showed stronger activity than bifonazole (Soković et al., 2006b). The antifungal activity of essential oil from Lavander showed therapeutic and antifungal potential during the 13-day observation period and cured the animals completely (Soković et al., 2007). The essential oil of thyme and its main component thymol was also tested for therapeutic potency in vivo. This oil completely cured animals infected with T. mentagrophytes within 24 days, with T. rubrum within 37 days, and with T. tonsurans within 32 days of treatment. The animals treated with the commercial drug bifonazole were cured after 14–15 days of treatment. Moreover thymol possessed higher therapeutic and antifungal activity than essential oil and cured the animals within 14 days (Figure 4.) (Soković et al., 2008c).

In vitro susceptibility of the turpentine oil obtained from Cluster pine Pinus pinaster oleoresin was evaluated against three Sudanese clinical isolates of Actinomadura madurae, which is the main causative agent of actinomycetoma in man and animals. The minimum inhibitory concentrations (MICs) of the oil ranged from 100.3-124.8 µl/ml, and the minimum microbicidal concentrations (MMCs) were between 100.3 µl/ml and 150.0 µl/ml. The main component of oil, α-pinene, exhibited prominent bioactivity with MICs ranging between 3.3 and

5.0 µl/ml, while the MMC was 10.0 µl/ml against the same clinical isolates. Cluster pine turpentine oil and α-pinene might be useful agents in the treatment of mycetoma caused by A. madurae (Figure 4.) (Stojković et al., 2008b). From the above results it can be concluded, that all essential oils tested showed beneficial antifungal activity both in vitro and in vivo.

After reviewing of the results of the antifungal activity of essential oils and individual components in vivo experiment, knowing that the composition of essential oils and the proportion of the tested individual components may be, to some extent, explain the differences between their activities. Menthol, camphor and thymol showed better antifungal activity than essential oils tested individually. Since the individual essential oils showed lower antifungal activity than the tested components, it is evident that the active principles can be explained by individual components. Although it is possible that interactions between the constituents of essential oils block the active principles of individual components when the treatment is the total essential oil. Added to that are antagonistic effect (Davidson & Parish, 1989) which does not mean that it can be completely neglected the role of individual components of essential oils on the expression of antifungal potential. Menthol, camphor and thymol, which showed the best antifungal activity in vivo among all tested components, are the dominant components of the essential oils of Spearmint and Thyme, and therefore can be justified by the high antifungal potential of these oils in vivo. The essential oil of Sweet basil, which is known for the beneficial activity of the skin, healing wounds, etc. is used to treat fungal infections showed good antifungal activity, but only better than lemon and orange. Dominant component of this oils was linalool, which proved to be good, but the tested components as one of the weaker fungicides, in front of limonene. Similarly, Lemon and Bitter orange oil showed the lowest antifungal potential, as well as among individual components of limonene, which is present in these oils with a high proportion, which is certainly influenced the decrease in the efficiency of oil. Essential oils of S. officinalis and L. angustifolia have proved to be most effective in the treatment of experimental induced dermatomycoses. If we compare the results obtained during investigation of antifungal activity of essential oils in vitro, and this generated in vivo, it is obvious that the essential oil of Sage and Lavander reacted with lower potential in vitro. In vivo experiments, these oils, in contrast, have proved to be most effective. Obviously, they have better therapeutic activity than other essential oils. In addition, it is known that Sage, Lavender and above all, always been used to treat various skin diseases and cosmetic products for skin care (Bremnes, 1994). Lavender essential oil possessed as the dominant components linalool, linalyl acetate, limonene, cineole and camphor. Good efficacy of essential oil it can be explained by interactions of individual components, but given the importance of some of the components, especially interactions linalyl acetate and linalool (Lis-Balchin et al., 1998). These essential oils and components could represent possible alternatives for the treatment of animals and humans infected by dermatomycetes. However, there are still only limited data available on the antifungal activity of essential oils towards human fungal pathogens in vivo (Soković, 2001).

Extracts of seventeen microfungi (Alternaria alternata, Cladosporium cladosporioides, C. fulvum, Fusarium sporotrichioides, F. trincintum, Paecilomyces variotii, Penicillium ochrochloron, P. funiculosum, Phoma magdonaldii, Phomopsis helianthi, Stachybotrys chartarum, Trichoderma viride, and five dermatomycetes, Epidermophyton floccosum, Microsporum canis, Trichophyton mentagrophytes, T. rubrum and T. tonsurans were tested against the yeast Candida albicans using a bioautographic assay on TLC plates. The extracts weer active against C. albicans at concrentations of 50-100 μg/mL. The extact of P. ochrochloron was most active. Further bioguided chemical analysis ofP. ochrochloron afforded two components with antimicrobial activity identified as (−)2,3,4-trihydroxybutanamid and (−)-erythritol. (−)-Erythritol showed moderate antifungal activity, while (−)2,3,4-trihydroxybutanamide was highly active against the fungi tested (Rančić et al., 2006). The antifungal activity of limonene was tested against Candida albicans and five dermatomycetes. The antifungal potential of limonene was evaluated against C. albicans using a bioautographic method on TLC plates. It showed better potential than bifonazole. When the activity of limonene towards five dermatomycetes was determined by the micro- and diffusion methods, it was more effective against these human and animal pathogens than bifonazole (Rančić et al., 2003). The antifungal activity of garlic bulb powder, allicin and the lozenge with 15% of garlic powder was tested using broth microdilution method against C. albicans. The tested garlic powder, as well the lozenge, have shown activity with MIC 1.25–7.50 mg/ml. The major compound, allicin, was highly active at a very low concentration with MIC 0.4 μg/ml. Those concentrations are lower then concentrations of commercially available fungicides (Kundaković et al., 2011). These antifungal agents in development offer extended half-lives, possibly reduced drug interaction profiles and good tolerance. In addition to activity against animal and human pathogens, they have a broad spectrum of activity including activity against resistant and emerging fungal species. According to these results microfungi may be the source of new biologically active substances, so special attention should be given to research on the biological activity of fungal metabolites and their application.

SUSCEPTIBILITY OF FUNGAL SPECIES TO TESTED COMPOUNDS

In our investigation, the extracts of different plants exhibited inhibitory effect on the growth of micromycetes. Among the tested extracts one of Phlomis fruticosa showed the best activity (10-20 μg/ml) (Table 1.). All investigated mosses extracts have been proved to be active against all fungi tested, where the ethanol extracts of silver moss showed the best potential (0.29-0.52 μg/ml) (Table 1). The results of the tested essential oils are summarized in Table 1. All tested essential oils exhibited antifungal activity ranging from 0.1-1250 μl/ml using different methods. Among all oils analyzed the essential oil of cinnamon was the most effective as an antifungal agent in concentration 0.1-1 μl/ml using microdilution method. Also, the components of essential oils such as carvacrol, thymol, menthol, showed very strong antifungal activity (0.02-300 μl/ml), where carvacol showed the best potential (0.1-0.5 μg/ml). Among all flavonoides, sesquiterpene lactones and other compounds analyzed in our investigation (Table 1), lactones from Centaurea pullata showed the best antifungal activity (0.0001-0.0007 μmol/ml). Thus, comparing the activity of our investigation (extracts, essential oils and components, pure compounds) we can conclude that the sesquiterpene lactones was the most effective as an antifungal agent, followed by essential oils and their components.

The human and food-borne pathogens are most frequently chosen for testing essential oil antimicrobial activity. Many laboratories deal with plant pathogens but fewer with animal pathogens. The essential oils which are the most tested compounds in our investigation of antifungal activity showed different effect on plant and human pathogens species. In our earlier investigation (Soković, 2001) essential oils in general exhibited higher antifungal activity against plant pathogen species (Phomopsis helianthi, Cladospoirium cladosporioides, Alternaria alternata, Fusarium species) than human (Trichophytonspecies, Microsporum cannis, Epidermophyton flocosum). The essential oils tested possessed different range of minimal fungicidal concentration where human pathogens were more resistant than plant pathogens: Mentha spicata 1-2 µl/ml for plant pathogens (pps) and 2 µl/ml for human pathogens (hps),M. piiperita 1.5-2.5 µl/ml pps and 2.5 µl/ml for hps, Thyme 0.125-0.25 µl/ml pps and 0.25 µl/ml for hps., Salvia species for pps 2-15 µl/ml and 3-20 µl/ml for hps, Lavandula species 0.5-9 µl/ml for pps and 0.5-10 µl/ml hps, Citrus aurantium 10 µl/ml for pps and 7-15 µl/ml for hps. Some oils had the same antifungal ability against both, pps and hps: Melissa officinalis (5-6 µl/ml), Rosmarinus officinalis (6-8 µl/ml), Citrus lemon (7-10 µl/ml), while Matricaria chamommilla exhibited better antifungal capacity against hps (6-9 µl/ml) than for pps (7-9 µl/ml) and Sweet basil (4-5 µl/ml for pps) and 3-5 µl/ml for hps. These results are confirmed by our investigation of antifungal activity with some other essential oils that we tested latter (Soković et al., 2002; Ristić et al., 2004; Simić et al., 2008;Džamić et al., 2009). Other researchers also obtained the similar results and confirmed that dermatomycetes are the most resistant, although numerous essential oils demonstrate high effectiveness against them. Among 22 samples of essential oils from 11 species of Cinnamomum, the oil of C. suvabenium was the most active against Microsporum canis, Trichophyton mentagrophytesand T. rubrum as well as some candidiasis (C. albicans and C. glabrata), (Kalemba & Kunicka, 2003). The results indicate that different essential oils have different efficacy. Also, the modes of action of essential oils are not the same against different fungal species. The mode of action of antimicrobial agents also depends on the type of fungal species and is mainly related to their cell wall structure and the outer membrane arrangement (Villar et al., 1986).

A NOVEL APPROACH TO SOLVE THE PROBLEM

Fungal disease is responsible for significant losses of global crop production every year, and thus has a major impact on the world's agricultural productivity. Numerous strategies have been developed in attempts to minimize the losses caused by plant pathogens. Traditional approaches are based on the avoidance of sources of infection, vector management, modification of cultural practices, the use of resistant varieties obtained through conventional breeding, cross protection and chemical control. While these methods have been successful in some cases, indeed there is a need for new approaches. Furthermore some fungicides are being withdrawn from the market because of their undesirable effects on the environment. Whereas this information is interesting with regards to economic aspects, such as the share of fungicides costs in the output or in the variable costs, the information is less useful with regards to environmental aspects: i.e., the amounts spent on fungicides do say little about the types and quantities used. Over recent decade, producers have used synthetic fungicides as the main tool to control this problem. It has been estimated that over 23 million kg of these synthetic fungicides are used annually worldwide and it is generally accepted that production and marketing of plants would be not possible without their use (Martinez-Romero et al., 2008). New strategies for disease control are therefore urgently required. The development of novel control strategies for plant diseases is particularly important for pathogens that are difficult to control using existing methods.

More recently the scientific community has turned its attention to secondary metabolites from actinobacteria and its exploitation for various purposes which include therapeutic, environmental and industrial applications. With developing microbial resistance and need for safe and cost-effective antifungal drugs, screening of some other source, i.e. micro- and macro-fungi, mosses, for potential bioactive secondary metabolites becomes necessary. Particularly desirable is the discovery of novel prototype therapeutic agents representing new chemical classes, that operate by different modes of action compared to existing agents. We were developed suitable approves during the last several years in order to find new solutions in the aim to discover new antifungal drugs either by testing already existing medical compounds, compounds from natural sources such as plants, micro- and macrofungi, or by combination of chemical compounds with natural one. Researchers also strive to elucidate the underlying biology of fungal microorganism both in vitro and in vivo.

The majority of our results are focused on investigation of antifungal activity of natural products isolated from plants, especially essential oils. We used conventional research methods for testing of antifungal activity and introduced some modification for the corresponding class of compounds and microorganisms in which we operate. For example, for testing of compounds in small quantity, we recommend using a microdilution method, and microathmosphere method for testing of volatile compounds. Bioautogaphic method on TLC plates is suitable for testing of plant and fungi extracts and fractions. Considering the fact that the fungi (micro and macro) may be the source of new biologically active substances, special attention in our laboratory is given to the research of biological activities of fungal metabolites and their application in protection and treatment of diseases caused by fungi and environmental protection. The broad diversity of the fungi, as well as their easy acquisition make them especially interesting for natural products screening program. The fungi possess high capacity of bio-synthesizing various metabolites possessing different structural and pharmacological characteristics. Many medicinal and therapeutic properties are attributed to the presence of active substances in fungi. Some such compounds are investigated because their known triggering mechanisms important for fungi, while other compounds are tested blindly for their antifungal properties.

Finally, research should be kept up in order to uncover as much potentially interesting data as possible, including toxicity against animal or human cells, mechanisms of action, effects in vivo, positive and negative interactions with common fungicides and so forth. Currently, these studies have produced a compounds suitable for the clinical trial stage. In summary, it is our belief that the study of plant, fungi and other natural sources as antimicrobial agents is necessary but the use of a appropriate method for investigation is essential. Finally, our results could be suggested for further clinical tests and for getting new information and possible application. To this should be given high priority.

CONCLUSION AND FUTURE TRENDS

Natural product-based fungicides are generally considered safer than synthetic herbicides, because of their relatively short environmental half-life and they are not harmful. The recent resurgence of interest in natural sources of bioactive compounds may, in part, be attributed to improved methods and instrumentation that has greatly reduced the time and effort required in natural product discovery programs. This interest is also associated with several other factors, including the realization that nature has already selected for very specific biological activities, that many natural compounds have yet to be discovered, and that the biological activities of relatively few of the known natural products have been characterized. All of the described and possible secondary metabolites have some kinds of inherent activity but in many cases these activities have not yet been discovered. Only the methods to detect their possible, perhaps until now unknown type of activity, has to be developed. There is no reason to suppose that the majority of the natural products including fungal metabolites should not exhibit some kind of biological function.

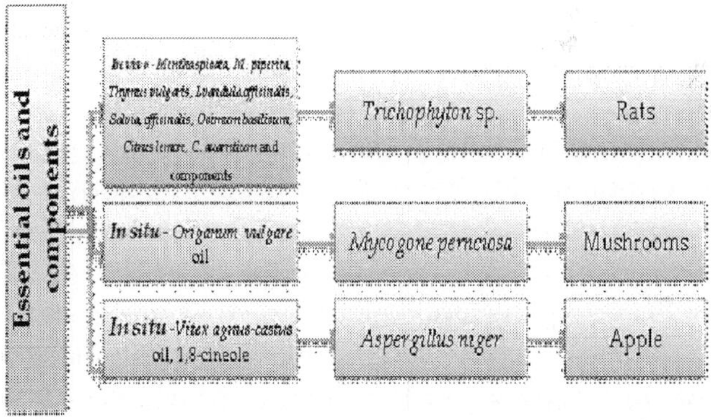

Figure 4: In situ and in vivo testing of antifungal activity of essential oils and components against pathogenic microfungi.

Hundreds of presently known bioactive metabolites originally was discovered as "inactive", natural product and their activity was only discovered later, investigating them with new more specific methods, or reisolated them (sometimes from different species). In our days, in fact, there would not be any reason to talk about "bioactive" or "inactive" secondary metabolites and treat them separately. Moreover, the study of natural products may lead to the discovery of novel target sites, and/or new classes of chemistry that can be developed for pathogen management. Structural diversity has been, and still remains, an invaluable source of lead compounds in developing novel products. A recent study on complementary synthetic and natural products confirmed that the later generally have higher molecular weights than the former. Such diversity may be useful to the synthetic chemist in developing new classes of fungicides. One indirect and important benefit of the chemical composition and structural characteristics of natural products (the absence of "unnatural" ring structures and the low content of heavy atoms) is that most of them are rapidly degraded in the natural environment to benign products. In addition to their structural features, natural products tend to have different target sites from conventional fungicides.

Our results contribute to the development of safe, effective, and inexpensive formulations and processes to reduce the presence of pathogens. The antifungal compounds identified by us as the most active against major pathogens are candidates for future studies of synergism, compatibility and activity in different systems. Isolation and identification of natural active components may include a multitude of different extractions, chemical modifications, and increase knowledge of their mechanisms of action. As essential part of obtaining natural fungicides is the development of bioassays. The development of fungal resistance to synthetic drugs poses a serious long-term trait to plant, animal and human health and environmental requirements. This could also possess as significant financial issues. The advantage of natural products compared to synthetic is not only in their non toxic characteristics but also in low costs. Growing of medicinal and aromatic plants and fungi is well established and in most cases economically justified. Identification and isolation of active components from plants and fungi is also good elaborated. Natural products with antifungal activity usually operate in very small concentrations, especially essential oils, and the for further application small amount are needed. All together makes them relative cheap and available as antifungal agents.

The future of fungicide management will probably be significantly influenced by research on natural products. Modern instrumentation has simplified the isolation and identification of lead compounds from which fungicides will be derived. The reviewed studies clearly demonstrate that natural products from plants and fungi present great potential for medical procedures and for the food, cosmetic, agricultural and pharmaceutical industries.

ACKNOWLEDGEMENTS

The authors are grateful to the Ministry of Education and Science of Serbia for financial support (Grant number 173032).

REFERENCES

1. K. Adam, A. Sivropoulu, S. Kokkini, T. Lanaras, M. Arsenakis, 1998Antifungal activities of Origanum vulgare subsp. hirtum, Mentha spicata, Lavandula angustifolia and Salvia fruticosa essential oils against human pathogenic fungi. Journal of Agricultural and Food Chemistry, 465173917450021-8561
2. K. Ando, S. Suzuki, T. Seaki, G. Tamura, K. Arima, 1969Funiculosin, a new antibiotic. Isolation, biological and chemical properties. The Journal of Antibiotics, 221891940021-8820
3. G. Alquini, E. R. Carbonero, 2004Polysaccharides from the fruit bodies of the basidiomycete L. sulphureus. FEMS Microbiology Letters, 230147521574-6968
4. T. M. Baratta, D. J. H. Dorman, G. S. Deans, C. A. Figueiredo, G. J. Barroso, G. Ruberto, 1998Antimicrobial and antioxidant properties of some commercial essential oils. Flavour and Fragranace Journal, 132352441099-1026
5. L. Barros, R. C. Calhelha, J. A. Vaz, I. C. F. R. Ferreira, P. Baptista, L. Estevinho, 2007Antimicrobial activity and bioactive compounds of Portuguese wild edible mushrooms methanolic extracts. European Food Research and Technology, 2251511561438-2377
6. J. Berdy, 1980CRC handbook of antibiotic compounds. CRC Press, 0-84933-450-0Raton, Florida, USA, 7.
7. J. Berdy, 2005Bioactive Microbial Metabolites, A Personal View. The Journal of Antibiotic, 5811260021-8820

8. D. C. Bishop, B. I. Thornton, 1997Evaluation of the antifungal activity of the essential oils of Monarda citridora var. citridora and Melaleuca alternifolia on post-harvest pathogens. Journal of Essential Oil Research, 977821041-2905
9. L. Bremnes, 1994Herbs. Eyewitness-Handbooks. DK Publishing, 1-56458-497-6York.
10. S. Brul, P. Coote, 1999Mode of action and microbial resistance mechanisms. Inernational Journal of Food Microbiology, 501170168-1605
11. L. B. Bulerman, Y. Lieu, A. S. Seier, 1977Inhibition of growth and aflatoxin production by cinnamon and clove oils, cinamicaldexyde and eugenol. Journal of Food Scence, 42110711091750-3841
12. J. M. Butler, A. W. Day, J. M. Henson, N. P. Money, 2001Pathogenic properties of fungal melanins. Mycologia, 93180027-5514
13. J. H. Clarke, 1966Studies on fungi in the root region. V. The antibiotic effects of root extracts of allium on some root surface fungi. Plant Soil, 2532400003-2079X
14. D. E. Conner, R. L. Beuchat, 1984Effects of essential oils from plants on growth of food spoilage yeasts. Journal of Food Scence, 494294341750-3841
15. M. M. Cowan, 1999Plant Products as Antimicrobial Agents. Clinical Microbiology Reviews, 1245645820893-8512
16. R. J. W. Cremlyn, 1991Agrochemicals: preparation and mode of action. John Wiley, 0-47127-669-3UK, 396.
17. A. Ćirić, 2010Antimikrobial activity of secondary metabolits izolated from Centaurea spruneri Boiss. & Heldr and Centaurea zuccariniana DC. Ph. D. Thesis, Faculty of Biology, University of Belgrade, Belgrade Serbia
18. A. Ćirić, A. Karioti, J. Glamoclija, M. Sokovic, H. Skaltsa, 2011Antimicrobial activity of secondary metabolites isolated from Centaurea spruneri Boiss. & Heldr. Journal oh the Serbian Chemical Society, 76127340352-5139
19. K. R. Daouk, M. S. Dagher, J. E. Sattout, 1995Antifungal activity of the essential oil of Origanum syriacum L. Journal of Food Protection, 58114711490036-2028X
20. Davidson P.M. & Parish E.M.1989Methods for testing the efficacy of food antimicrobials. Food Technology, 431481550015-6639

21. G. S. Deans, P. K. Svoboda, 1990The antimicrobial properties of Majoram (Origanum majoram L.) volatile oils. Flavour and Fragranace Journal, 5118711901099-1026
22. C. J. Delp, 1988Fungicide resistance in North America. APS Press, 978-0-89054-095-4St. Paul. Minesota, USA, 133.
23. S. Djeddi, A. Karioti, M. Soković, D. Stojković, R. Seridi, H. Skaltsa, 2007Minor Sesquiterpene Lactones from Centaurea pullata and Their Antimicrobial Activity. Journal of Natural Products, 7011179617990163-3864
24. S. Djeddi, A. Karioti, M. Soković, C. Koukoulitsa, H. Skaltsa, 2008A novel Sesquiterpene Lactones from Centaurea pullata: Structure elucidation, Antimicrobial Activity, and prediction of pharmacokinetic properties. Bioorganic & Medicinal Chemistry, 163115011610968-0896
25. A. Džamić, M. Soković, M. Ristić, S. Grujić-Jovanović, J. Vukojević, P. Marin, 2008Chemical Composition and Antifungal Activity of Salvia sclarea (Lamiaceae) Essential Oil. Archive of Biological Science, 6022332370354-4664
26. A. Džamić, M. Soković, M. Ristić, S. Grujić-Jovanović, J. Vukojević, P. Marin, 2009Chemical composition and antifungal activity of Illicium verum and Eugenia caryophyllata essential oils. Chemistry of Natural Compounds, 4522592610009-3130
27. A. Džamić, M. Soković, M. Ristić, M. Novaković, S. Grujić-Jovanović, V. Tešević, P. Marin, 2010Antifungal and antioxidant activity of Mentha longifolia (L.) Hudson (Lamiaceae) essential oil. Botanica Serbica, 34157611821-2158
28. O. S. Duke, E. F. Dayan, G. J. Romagni, M. A. Rimando, 2000Natural products as sources of herbicides: current status and future trends. Weed Research, 40991111365-3180
29. S. C. Garg, N. Siddiqui, 1992Antifungal activity of some essential oil isolates. Pharmazie, 474674680031-7144
30. M. A. Giudici, C. M. Regente, L. de la Canal, 2000A potent antifungal protein from Helianthus annuus flowers is a trypsin inhibitor. Plant Physiology and Biochemistry, 388818880981-9428
31. J. Glamočlija, 2006aPrevention and treatment of Mycogone perniciosa Magn. in mushroom units of Agaricus bisporus (J.

Lange) Imbach. Ph.D.Tthesis Faculty of Biology, University of Belgrade, Belgrade, Serbia

32. J. Glamočlija, M. Soković, J. Vukojević, I. Milenković, L. J. L. D. Van Griensven, 2006bChemical Composition and Antifungal Activities of Essential Oils of Satureja thymbra L. and Salvia pomifera ssp. Calycina (Sm.) Hayek. Journal of Essential Oil Research, 181151171041-2905
33. J. Glamočlija, M. Soković, J. Vukojević, I. Milenković, V. Tešević, L. J. L. D. Van Griensven, 2007Effect of Oregano essential oil on infection of Agaricus bisporus by Mycogone perniciosa in vitro and in the Mushroom growing unit. International Journal of Medicinal Mushrooms, 93-43001521-9437
34. J. Glamočlija, M. Soković, D. Grubišić, J. Vukojević, I. Milenković, M. Ristić, 2009Antifungal activity of Critmum maritimum L. essential oil and its components against mushroom pathogen Mycogone perniciosa Mang. Chemistry of Natural Compounds, 45196970009-3130
35. J. Glamočlija, M. Soković, J. Šiljegovic, M. Ristić, A. Ćirić, D. Grubišić, 2011aChemical Composition and Antimicrobial Activity of Echinophora spinosa L. (Apiaceae) Essential Oil. Records of Natural Products, 543193230000-1307
36. J. Glamočlija, M. Soković, V. Tešević, G. A. Linde, Colauto. N. Barros, 2011bChemical characterization of Lippia alba essential oil: an alternative to control green molds. Brazilian Journal of Microbiology, 42153715461517-8382
37. G. S. Griffin, G. S. Wylie, L. J. Markham, N. D. Leach, 1999The role of structure and molecular properties of terpenoids in determining their antimicrobial activity. Flavour and Fragranace Journal, 143223321099-1026
38. N. Guterson, 1990Microbial fungicides: recent approches to elucidating mechanisms. Critical Reviews in Biotechnology, 1069910738-8551
39. H. Hanel, W. Raether, 1988A more sophisticated method of determining the fungicidal effect of water-insoluble preparations with a cell harvester, using miconazole as an example. Mycoses, 3131481541439-0507

40. A. L. Harvey, 2008Natural products in drug discovery. Drug Discovery Today, 131989490113 59-6446
41. G. Herger, F. Klingauf, D. Mangold, E. H. Pommer, M. Scheter, 1988Die Wirkung von Auszigen aus dem Sachalin-Staudenknoterich Reynoutria suchalinensis (F. Schmidt) Nakai gegen Plizkrankheiten, insbesondere Echte Mehltauplize. Nachrchtenblatt der Deutsche Pflanzenschutzdienst, 4056600027-7479
42. I. A. Hussain, F. Anwar, P. S. Nigam, S. D. Sarker, J. E. Moore, J. R. Rao, A. Mazumdar, 2011Antibacterial activity of some Lamiaceae essential oils using resazurin as an indicator of cell growth. Food Science and Technology, 44119912060000-0023
43. D. Ilić, V. Nikolić, A. Ćirić, M. Soković, T. Stanojković, T. Kundaković, M. Stanković, Lj. Nikolić, 2012Cytotoxicity and antimicrobial activity of allicin and its transformation products. Journal of Medicinal Plants Research, 61596511996-0875
44. H. Ishii, 1995Monitoring of fungicide resistance in fungi: biological to biochemical approaches. In: Molecular Methods in Plant Pathology, Singh, S.U., Singh, P.R., (Eds.), 4834950-87371-877-1Publisher, Boca Ratton, London, Tokyo
45. K. Isono, 1990Antibiotics as noon polution agricultural pesticides. Comments on Agricultural and Food Chemistry, 21231420892-2101
46. P. Janaćković, V. Tešević, P. D. Marin, S. M. Milosavljević, B. Petković, M. Soković, 2002Polyacetylenes and a sesquiterpene lactone from Ptilostemon strictus. Biochemical Systematics and Ecology, 3069710000-0305
47. P. Janaćković, M. Sokovic, Lj. Vujisic, V. Vajs, I. Vuckovic, Z. Krivosej, P. D. Marin, 2011Composition and Antimicrobial Activity of Seseli globiferum Essential Oil. Natural Product Communacations, 68116311660193-4578X
48. A. M. Janssen, J. J. C. Scheffer, Svendsen. A. Baerheim, 1987Antimicrobial activity of essential oils: a 1976-1986 literature review. Aspect of the test methods. Planta Medica,533953980032-0943
49. A. Jaspers, 1994Mode of action of the phenylpyrole fungicide fenpiclonil in Fusarium sulphureum. Ph. D. Thesis, Wageningen, Holland.

50. J. J. Kabara, 1991Phenols and Chelators, In: Food Preservatives, Russell, N.J., Gould, G.W., (Eds.), 200214ISBN 030647736X, 9780306477362, Blackie: London, UK
51. D. Kalemba, A. Kunicka, 2003Antibacterial and antifungal properties of essential oils. Current Medicinal Chemistry, 108138290929-8673
52. A. Karioti, M. Soković, A. Ćirić, C. Koukoulitsa, R. A. Bilia, H. Skaltsa, 2011Antimicrobial Properties of Quercus ilex L. Proanthocyanidin Dimers and Simple Phenolics: Evaluation of Their Synergistic Activity with Conventional Antimicrobials and Prediction of Their Pharmacokinetic Profile. Journal of Agricultural and Food Chemistry, 5912641264220021-8561
53. K. Knobloch, A. Pauli, B. Iberl, N. Weis, H. Weigand, 1988Mode of action of essential oil components on whole cells of bacteria and fungi in plate tests, In: Bioflavor '87, Schreier, P. (ed.), 287299Walter de Gruyter Verlag, 0-89925-290-7New York
54. M. Konstantinopoulou, A. Karioti, S. Skaltsas, H. Skaltsa, 2003Sesquiterpene lactones from Anthemis altissima and their anti-Helicobacter activity. Journal of Natural Products, 6656997020163-3864
55. E. Koukoulitsa, D. George, D. E. Geromichalos, H. Skaltsa, 2005VolSurf analysis of pharmacokinetic properties for several antifungal sesquiterpenelactones isolated from Greek Centaurea sp. Journal of Computer-Aided Molecular Design, 1986176230092-0654X
56. M. Krauze-Baranowska, W. Cisowkski, M. Wiwart, B. Madziar, 1999Antifungal biflavones from Cupressocyparis leylandi. Planta Medica, 655725730032-0943
57. J. Kukić, V. Popović, S. Petrović, P. Mucaji, A. Ćirić, D. Stojković, M. Soković, 2008Antioxidant and antimicrobial activity of Cynara cardunculus extarcts. Food Chemistry, 1078618680308-8146
58. D. T. Kundaković, D. A. Ćirić, D. M. Soković, T. M. Milenković, D. V. Nikolić, S. G. Nikolić, 2011Antimicrobial activity of lozenge with garlic bulb powder. Hemijska industrija, 6556076100036-7598X
59. I.. Kurobane, R. Hutchinson, L. Vining, 1981The biosynthesis of fulvic acid, a fungal metabolite of heptaketide origin. Tetrahedron Letters, 224934960040-4039

60. D. T. Lakshmipathy, K. Kannabiran, 2010Review on dermatomycosis: pathogenesis and treatment. Natural Sciences, 277267312150-4091
61. R. J. Lambert, P. N. Skandamis, P. J. Coote, G. J. Nychas, 2001A study of the minimum inhibitory concentration and mode of action of oregano essential oil, thymol and carvacrol. Journal of Applied Microbiology, 914534621365-2672
62. M. Lis-Balchin, G. S. Deans, E. Eaglesham, 1998Relationship between bioactivity and chemical composition of commercial essential oils. Flavour and Fragranace Journal, 13981041099-1026
63. Grbić. M. Ljaljević, J. Vukojević, M. D. Soković, D. Grubišić, M. Ristić, 2007The effect of Nepeta rtanjensis essential oil on test micromycetes mycelia growth. Proceedings Natural Science Matica Srpska, Novi Sad, 1131731770352-4906
64. Grbić. M. Ljaljević, J. Vukojević, M. Stupar, D. Grubišić, 2011In vitro antifungal and demelanizing activity of Nepeta rtanjensis essential oil agaisnt the human pathogen Bypolaris spicifera. Archives of Biological Sciences, 6338979050354-4664
65. J. W. Mansfield, 1983Antimicrobial compounds, In: Biochemical Plant Pathology, J. A. Callow, (Ed.), 237265Jonh Wiley and Soons, 100471900923UK
66. B. Marinković, P. D. Marin, J. Knežević-Vukčević, M. D. Soković, D. Brkić, 2002Activity of Essential Oils ofThree Micromera species (Lamiaceae) Against Micromycetes and Bacteria. Phytoterapy Research, 163363390095-1418X
67. T. Marković, P. Chatzopoulou, J. Šiljegović, M. Nikolić, J. Glamočlija, A. Ćirić, M. Soković, 2011Chemical Analysis and Antimicrobial Activities of the Essential Oils of Satureja thymbra L. and Thymbra spicata L. and their main components. Archives of Biological Sciences, 6324574640354-4664
68. J. C. Maruzella, J. Balter, 1959The action of essential oils on phytopathogenicfungi. PlantDiseaseReporter, 43114311470032-0811
69. D. Martínez-Romero, M. Serrano, G. Bailén, F. Guillén, P. J. Zapata, J. M. Valverde, S. Castillo, M. Fuentes, D. Valero, 2008The use of a natural fungicide as an alternative to preharvest

synthetic fungicide treatments to control lettuce deterioration during postharvest storage. Postharvest Biology and Technology, 47154600925-5214

70. R. P. Maskey, I. Grun-Wollny, H. Laatsch, 2003Isolation, structure elucidation and biological activity of 8-O-methylaverufin and 1, 8-Odimethylaverantin as new antifungal agents from Penicillium chrysogenum. The Journal of Antibiotics, 564884910021-8820
71. Y. Masaoka, M. Kojima, S. Sagihara, T. Yohohara, M. Koshino, A. Ischihara, 1993Dissolution of ferric phoshate by alfaalfa (Medicago sativa L.) root exudates. Plant Soil, 15575780003-2079X
72. T. Matshumoto, S. Tahara, 2001Ampelopsin, a major antifungal constituten fro Salix sachalinensis, and its methyl ethers. Nippon Kagacu Kaishi, 786586670369-4577
73. N. Mimica-Dukić, S. Kujundžič, M. Soković, M. Couladis, 2003Essential Oil Composition and Antifungal Activity of Foeniceum vulgare Mill. Obtained by Different Distillation Conditions., Phytoterapy Research, 173683710095-1418X
74. S. M. Milosavljević, V. Tešević, I. Vučković, M. Jadranin, V. Vajs, M. Soković, P. Janaćković, A. Jovanović, 2007Composition and antifungal activity of the essential oil of Seseli annuum wild growing in Serbia. Fitoterapia, 7843193220036-7326X
75. A. K. Mishra, M. K. Dubey, 1994Evaluation of some essential oils for their toxicity against fungi causing deterioration of stored food commodities. Applied and Environmental Microbiology, 60110111050099-2240
76. R. Montes-Belmont, M. Carvajal, 1998Control of Aspergillus flavus in maize with plant essential oils and their components. Journal of Food Protection, 36166190036-2028X
77. A. C. P. Moreira, E. O. Lima, P. A. Wanderley, E. S. Carmo, E. L. de Souza, 2010Chemical composition and antifungal activity of Hyptis suaveolens (L.) Poit leaves essential oil against Aspergillus species. Brazilian Journal of Microbioliology 4128331517-8382
78. M. J. D. Moretti, T. A. Peana, A. Franceschini, C. Carta, 1998In vivo activity of Salvia officinalis oil against Botrytis cinerea. Journal of Essential Oil Research, 101571601041-2905

79. K. P. Mukherjee, K. Saha, P. B. Saha, M. Pal, J. Das, 1996Antifungal activities of the leaf extract of Cassia tora Linn. (Fam. Leguminosae). Phytoterapy Research, 105215220095-1418X
80. F. Müller-Riebau, B. Berger, O. Yegen, 1995Chemical composition and fungitoxic properties to phytopathogenic fungi of essential oils of selected aromatic plants growing wild in Tukey. Journal of Agricultural and Food Chemistry, 43822622266002 1-8561
81. Y. J. Nam, K. H. Kim, Y. J. Know, Y. M. Han, H. K. Son, C. U. Lee, D. J. Choi, M. B. Know, 2000O-Methylsclererotiorinamine, Antagonist of the Grb2-SH2 Domain, Isolation from Penicillium multicolor. Journal of Natural Products, 63130313050163-3864
82. S. M. A. Naqvi, S. K. Chauhan, 1980Effect of root exudates on the spore germination of rhizosphere and rhizoplane mycoflora of chili (Capsicum annuum L.) cultivars. Plant Soil, 553974020003-2079X
83. J. Nestorović, D. Mišić, B. Šiler, M. Soković, J. Glamočlija, A. Ćirić, V. Maksimović, D. Grubišić, 2010Nepetalactone content in shoot cultures of three endemic Nepeta species and the evaluation of their antimicrobial activity. Fitoterapia, 816216260036-7326X
84. J. D. Nosanchuk, A. Casadevall, 2006Impact of melanin on microbial virulence and clinical resistance to antimicrobial compounds. Antimicrobial Agents and Chemotherapy, 50351935280066-4804
85. Oppenheimer Wolff & Donnelly1997Possibilities for Future E.U. Environmental Policy on Plant Protection Products: Synthesis Report. Oppenheimer Wolff & Donnelly, Brussels, Available from http://ec.europa.eu/environment/pps/Synthesis Report
86. A. E. Osbourn, 1996Saponins and plant defence- A soap story. Trends in Plant Science, 1481360-1385
87. M. Ozcan, N. Boyraz, 2000Antifungal properties of some herb decoctions. European Food Research and Technology, 21286881438-2377
88. T. Pacher, M. Bacher, O. Hofer, H. Greger, 2001Stress induced carbazole phytoalexins in Glycosmis species. Phytochemistry, 5811291350031-9422
89. V. C. Pawar, V. S. Thaker, 2006In vitro efficacy of 75 essential oils against Aspergillus niger. Mycoses, 493163231439-0507

90. B. Pejin, A. Sabovljević, M. Soković, J. Glamočlija, A. Ćirić, M. Vujičić, M. Sabovljević, 2012Antimicrobial activity of Rhodobryum ontariense. Hemijska industrija, Research note, in press, 0036-75980367598X
91. A. K. Picman, F. E. Schneinder, J. Piceman, 1995Effect of flavonoides on mycelial growth of Verticillium albo-atrum. Biochemical Systematics and Ecology, 236836930305-1978
92. I. P. Raoha, S. Remes, M. Hienomen, M. Hopia, M. Kibkonen, T. Kajada, K. Pihlaja, H. Vuorela, P. Vuorela, 2000Antimicrobial effect of Finissh plant extract containing flavonoids and other phenolic compounds. International Journal of Food Microbiology, 563120168-1605
93. A. Rančić, M. Soković, L. Van Griensven, J. Vukojević, D. Brkić, M. Ristić, 2003Antimicrobial Activity of Limonene. Matieres Medicales, 2383880455-6224
94. A. Rančić, 2004Antimicrobial efefct of metabolites isolated from Peniscillium ochrochloron Biorge, Master Thesis, Faculty of Biology, University of Belgrade, Belgrade, Serbia
95. A. Rančić, M. Soković, J. Vukojević, A. Simić, P. Marin, S. Duletić-Laušević, D. Đoković, 2005Chemical Composition and Antimicrobial Activities of Essential Oils of M.odorata (L.) Scop, Hypericum perforatum L and Helichrysum arenarium (L.) Moench. Journal of Essential Oil Research, 1743413451041-2905
96. A. Rančić, M. Soković, A. Karioti, J. Vukojević, H. Skaltsa, 2006Isolation and structural elucidation of two secondary metabolites from the filamentous fungus Penicillium ochrochloron with antimicrobial activity. Environmental Toxicology and Pharmacology, 22180841382-6689
97. J. L. Rios, M. C. Recico, 2005Medicinal plants and antimicrobial activity. Journal of Ethnopharmacology, 10080840378-8741
98. B. V. M. Reddy, P. Angers, A. Gosselin, J. Arul, 1998Characterization and use of essential oil from Thymus vulgaris against B. cinerea and Rh. stolonifera in strawberry fruits. Phytochemistry, 4715151520031-9422
99. M. D. Ristić, S. Duletić-Laušević, J. Knežević-Vukčević, P. D. Marin, D. Simić, J. Vukojević, P. Janaćković, V. Vajs, 2000Antimicrobial

activity of essential oils and ethanol extract of Phlomis fruticosa L. (Lamiaceae). Phytoterapy Research, 142672710095-1418X

100. A. Sabovljević, M. Soković, M. Sabovljević, D. Grubišić, 2006Antimicrobial Activity of Bryum argenteum. Fitoterapia, 771441450036-7326X
101. A. Sabovljević, M. Soković, J. Glamočlija, A. Ćirić, M. Vujičić, B. Pejin, M. Sabovljević, 2011Bio-activities of extracts from some axenically farmed and naturally grown bryophytes. Journal of Medicinal Plants Research, 545655711996-0875
102. V. Saroglou, A. Karioti, A. Rančić, K. Dimas, C. Koukoulitsa, M. Zervou, H. Skaltsa, 2010Sesquiterpene Lactones from Anthemis melanolepis and Their Antibacterial and Cytotoxic Activities. Prediction of Their Pharmacokinetic Profile. Journal of Natural Products, 73222422460163-3864
103. J. Sato, K. Goto, F. Nanjo, S. Kawai, K. Murata, 2000Antifungal activity of plant extracts against Arthirnium sacchari and Chaetonium funicola. The Journal of Bioscience and Bioengineering, 904424461389-1723
104. S. Schimtz, M. Weidenbörner, B. Kunz, 1993Herbs and spice as selective inhibitors of mould growth. Chemie, Microbiologie, Technologie der Lebensmittel, 151751770366-7154
105. N. Sharma, A. Tripathi, 2008Effect of Citrus sinensis (L.) Osbeck epicarp essential oil on growth and morphogenesis of Aspergillus niger (L.) Van Tieghem. Microbiological Research, 1633373440953-7562
106. A. L. Shelef, 1983Antimicrobial effects of spice. Journal of Food Safety, 629441745-4565
107. S. Shin, S. Lim, 2004Antifungal effects of herbal essential oils alone and in combination with ketoconazole against Trichophyton spp. Journal of Applied Microbiology, 97128912961365-2672
108. M. Shimoni, E. Putiewsky, U. Ravid, R. Reuveni, 1993Antifungal activity of volatile fractions of essential oils from four aromatic wild plants in Israel. Journal of Chemical Ecology, 19112911330098-0331
109. N. C. C. Silva, J. A. Fernandes, 2010Biological properies of medicinal plants: A Review of Thier Antimicrobial Activity.

Journal of Venomous Animals and Toxins including Tropical Diseases, 164024131678-9199

110. S. M. A. Silva, M. Weidenbörner, A. S. J. Cavaleiro, 1998Growth control of different Fusarium species by selected flavones and flavonid mixtures. Mycological Research, 1026386400953-7562
111. P. D. Singh, J. H. Johnson, C. A. Aklonis, J. O'Sullivan, 1986SQ 30,957, a new antibiotic produced by Penicillium funiculosum. The Journal of Antibiotics, 39105410580021-8820
112. J. Singh, N. N. Tripathi, 1999Inhibition of storage fungi of blackrgram (Vigna mungo L.) by some essential oils. Flavour and Fragranace Journal, 14141099-1026
113. A. Simić, M. D. Soković, M. Ristić, S. Grujić-Jovanović, J. Vukojević, P. D. Marin, 2004The chemical composition of some Lauraceae essential oils and their antifungal activities. Phytoterapy Research, 1897137170095-1418X
114. A. Simić, A. Rančić, M. D. Soković, M. Ristić, S. Grujić-Jovanović, J. Vukojević, P. D. Marin, 2008Essential Oil Composition of C.ymbopogon winterianus and Carum carvi and Their Antimicrobial Activities. Pharmaceutical Biology, 4664374411388-0209
115. H. Skaltsa, D. Lazari, C. Panagouleas, E. Georgiadou, B. Garcia, M. Soković, 2000aSesquiterpene lactones from Centaurea thessala and Centaurea attica. Antifungal activity. Phytochemistry, 559039080031-9422
116. H. Skaltsa, D. Lazari, B. Garcia, R. J. Pedro, M. Soković, T. Constantinidis, 2000bSesquiterpene lactones from Centaurea achaia, a greek endemic species. Antifungal activity. Zeitschrift für Naturforschung, 55c5345390939-5075
117. H. Skaltsa, C. Demetzos, D. Lazari, M. Soković, 2003Essential oil analysis and antimicrobial activity of eight Stachys species from Greece. Phytochemistry, 6437437520031-9422
118. P. N. Skandamis, J. G. Nychas, 2000Development and evaluation of a model predicting the survival of Escherichia coli O157:H7 NCTC 12900 in homemade eggplant salad at various temperatures, pHs, and oregano essential oil concentrations. Applied and Environmental Microbiology, 66164616530099-2240

119. F. R. Smiderle, L. G. Sassaki, J. van Arkel, M. Iacomini, J. H. Wichers, L. J. L. D. Van Griensven, 2010High Molecular Weight Glucan of the Culinary Medicinal Mushroom Agaricus bisporus is an α-Glucan that Forms Complexes with Low Molecular Weight Galactan. Molecules, 15581858301420-3049
120. A. Sokmen, M. B. Jones, M. Erturk, 1999The in vivo antibacterial activity of Turkish medicinal plants. The Journal of Ethnopharmacology, 6779860378-8741
121. M. D. Soković, P. D. Marin, D. Brkić, 2000Antifungal activity of ethanolic extract of Phlomis fruticosa L. Archives of Biological Sciences, 522032080354-4664
122. D. M. Soković, 2001Antifungal activities of essential oil of selected aromatic and medicinal plants in vitro and in vivo, Ph.D. Thesis, Faculty of Biology, University of Belgrade, Yugoslavia
123. M. Soković, O. Tzakou, D. Pitarokili, M. Couladis, 2002Antifungal activities of seleted aromatic plants growing wild in Greece. Food Nahrung, 53173200000-0027X
124. M. D. Soković, M. Ristić, D. Grubišić, 2004Chemical Composition and Antifungal Activity of the Essential Oil from Juniperus excelsa Berries. Pharmaceutical Biology, 424-53283311388-0209
125. M. D. Soković, D. Grubišić, M. Ristić, 2005Chemical Composition and Antifungal Activity of the Essential Oils from Leaves, Calyx and Corolla of Salvia brachyodon Vandas. Journal of Essential Oil Research, 1722272291041-2905
126. M. D. Soković, L. J. L. D. Van Griensven, 2006aAntimicrobial activity of essential oils and their components against the three major pathogens of the cultivated button mushroom, Agaricus bisporus. The European Journal of Plant Pathology, 11632112241573-8469
127. M. D. Soković, J. Glamočlija, P. D. Marin, D. Brkić, J. Vukojević, D. Jovanović, N. Bulajić, D. Kataranovski, 2006bAntifungal Activity of the Essential Oil of Mentha x piperita. Pharmaceutical Biology, 445115151388-0209
128. M. D. Soković, J. Glamočlija, P. D. Marin, D. Brkić, J. Vukojević, D. Jovanović, N. Bulajić, D. Kataranovski, 2007Experimentally Induced Dermatomycoses at Rats and Treatment with Lavandula angustifolia Essential Oil. Proceedings Natural Science Matica Srpska, Novi Sad, 1132492540352-4906

129. M. Soković, P. Marin, J. Vukojević, M. Ristić, J. Glamočlija, A. Ćirić, 2008aChemical composition of Hyssopus officinalis L. essential oil and its antifungal activity. Matieres Medicales, 2641500455-6224

130. M. Soković, J. Glamočlija, A. Ćirić, D. Grubišić, D. Stojković, M. Ristić, 2008bAntimicrobial activity of essential oils isolated from different parts of endemic plan Portenschlagiella ramosissima Tutin. Journal of Essential Oil Research, 2043693721041-2905

131. M. Soković, J. Glamočlija, A. Ćirić, D. Kataranovski, P. Marin, J. Vukojević, D. Brkić, 2008cAntifungal activity of the essential oil of Thymus vulgaris L. and thymol on experimentaly induced dermatomycoses. Drug development and industrial pharmacy, 3412138813930363-9045

132. M. Soković, D. Brkić, A. Džamić, M. Ristić, P. D. Marin, 2009aChemical composition and antifungal activity of Salvia desoleana Atzei & Picci essential oil and its major components. Flavour and Fragranace Journal, 24283871099-1026

133. M. Soković, J. Vukojević, P. D. Marin, D. D. Brkić, V. Vajs, L. J. L. D. van Griensven, 2009bChemical Composition of Essential Oils of Thymus and Mentha Species and Their Antifungal Activities. Molecules, 1412382491420-3049

134. M. Soković, D. Stojković, J. Glamočlija, A. Ćirić, M. Ristić, D. Grubišić, 2009cSusceptibility of pathogenic bacteria and fungi to essential oils of wild Daucus carota. Pharmaceutical Biology, 47138431388-0209

135. D. Stojković, J. Glamočlija, M. Soković, D. Grubišić, S. Petrović, J. Kukić, M. Ristić, 2008aChemical composition, antimicrobial and antiradical properties of the essential oils of Seseli globiferum fruits. Natural Product Communacations, 311193519380193-4578X

136. D. Stojković, M. Soković, J. Glamočlija, A. Džamić, M. Ristić, A. Fahal, S. Khalid, I. Đujić, S. Petrović, 2008bSusceptibility of three clinical isolates of Actinomodura madurae to α-pinene, the bioactive agent of Pinus pinaster turpentine oil. Archives of Biological Sciences, 6046977010354-4664

137. S. Stojković, S. Petrović, J. Kukić, A. Dzamić, M. Ristić, M. Milenkovic, J. Glamočlija, M. Soković, D. Stojković, 2009Chemical composition and antimicrobial and antioxidant

activity of Seseli rigidum flower essential oil. Chemistry of Natural Compounds, 4522532560009-3130

138. D. Stojković, J. Šiljegović, M. Nikolić, A. Ćirić, J. Glamočlija, M. Soković, I. Milenković, 2011aSulhpur Polypore, Laethiporus sulphureus extract as natural preservative for the in vivo control of Aspergillus flavus in tomato paste. The 6th International Medicinal Mushroom Confrence, Book of Abstract, 6364Zagreb, Croatia, Sep 25-29, 2011
139. D. Stojković, M. Soković, J. Glamočlija, A. Džamić, A. Ćirić, M. Ristić, D. Grubišić, 2011bChemical composition and antimicrobial activity of Vitex agnus-castus L. fruits and leaves essential oils. Food Chemistry, 128101710220308-8146
140. B. Šiler, D. Mišić, J. Nestorović, T. Banjanac, J. Glamočlija, M. Soković, A. Ćirić, 2010Antibacterial and Antifungal Screening of Centaurium Pulchellum Crude Extracts and main Secoiridoid Compounds. Natural Product Communacations, 51015251530019 3-4578X
141. J. Šiljegović, D. Stojković, M. Nikolić, J. Glamočlija, M. Soković, A. Ćirić, 2011aAntimicrobial Activity of Aqueous extract of Laetiporus sulphureus (Bull.; Fr.) Muril. Proceedings Natural Science Matica Srpska, Novi Sad, 1202993050352-4906
142. J. Šiljegović, J. Glamočlija, M. Soković, I. Vučković, V. Tešević, S. Milosavljević, D. Stešević, 2011bComposition and Antimicrobial Activity of Seseli montanum subsp. tommasinii Essential Oil. Natural Product Communacations, 622632660193-4578X
143. R. Theodori, A. Karioti, A. Rančić, H. Skaltsa, 2006Linear Sesquiterpene Lactones from Anthemis auriculata and Their Antibacterial Activity. Journal of Natural Products, 6946626640163-3864
144. D. P. Thomson, 1989Fungitoxic activity of essential oil components on food storage fungi. Mycologia, 811511530027-5514
145. B. Trillini, R. E. Velasquez, R. Pellegrino, 1996Chemical composition and antimicrobial activity of essential oil of Piper angustifolium. Planta Medica, 623723730032-0943
146. A. Turkoglu, M. E. Durau, N. Mercan, I. Kivrak, K. Gezer, 2007Antioxidant and antimicrobial activities of L. sulphureus (Bull) Murill. Food Chemistry, 1012672730308-8146

147. J. Ushiki, Y. Hayakawa, T. Tadano, 1996Medicinal plants for suppressing soil-borne plant diseases I. Screening for medicinal plant with antimicrobial activity in root. Journal of Soil Science and Plant Nutrition, 424234261581-9175
148. J. Ushiki, S. Tahara, Y. Hayakawa, T. Tadano, 1997Medicinal plants for suppressing soil-borne plant diseases. Journal of Soil Science and Plant Nutrition, 441571651581-9175
149. V. Vajs, N. Todorović, M. D. Ristić, V. Tešević, B. Todorović, P. D. Marin, S. Milosavljević, 1999Guaianolides from Centaurea nicolai; Antifungal activity. Phytochemistry, 523833860031-9422
150. V. Vajs, S. Trifunovic, P. Janackovic, M. Sokovic, S. Milosavljevic, V. Tesevic, 2004Antifungal Activity of davananone-type sesquiterpenes from Artemisia lobelii var. Conescenes. Journal oh the Serbian Chemical Society, 69119699720352-5139
151. A. J. Vaz, L. Barros, A. Martins, C. Santos-Buelga, M. H. Vasconcelos, I. C. F. R. Ferreira, 2011Chemical composition of wild edible mushrooms and antioxidant properties of their water soluble polysaccharidic and ethanolic fractions. Food Chemistry, 1266106160308-8146
152. M. Veljić, M. Tarbuk, D. M. Petar, A. Ćirić, M. Soković, M. Marin, 2008Antimicrobial Activity of Methanol Extracts of Some Genuine Mosses from Serbia. Pharmaceutical Biology, 46128718751388-0209
153. M. Veljić, A. Djurić, M. Soković, A. Ćirić, J. Glamočlija, P. Marin, 2009Antimicrobial activity of mathanol extracts of Fontinalis antipyretica, Hypnum cupressiforme and Ctenidium molluscum. Archives of Biological Sciences, 6122252290354-4664
154. A. Villar, J. L. Rios, M. C. Recio, D. Cortes, A. Cave, 1986Antimicrobial activity of alkaloids. II. Relation between chemical composition and antimicrobial activity. Planta Medica, 65565570032-0943
155. K. B. Vivek, I. Y. Jung, C. K. Sun, 2009Antifungal potential of essential oil and various organic extracts of Nandina domestica Thunb. against skin infectious fungal pathogens. Applied Microbiology and Biotechnology, 83112711330175-7598

156. R. W. S. Weber, A. Mucci, P. Davoli, 2003Laetiporic acid, a new polyene pigment from the wood-rotting basidiomycete L. sulphureus (Polyporales, Fungi). Tetrahedron letters, 45107510780040-4039
157. E. D. Wedge, G. C. J. Galindo, A. F. Macias, 2000Fungicidal activity of natural and synthetic sesquiterpene lacton analogs. Phytochemistry, 537477570031-9422
158. M. Weidenbörner, H. Hindorf, C. H. Jha, P. Tsotsonos, H. Egge, 1989Antifungal activity of isoflavonoids against storage fungi of the genus Aspergillus. Phytochemistry, 12331733190031-9422
159. M. Weidenbörner, H. Hindorf, C. H. Jha, P. Tsotsonos, H. Egge, 1990Antifungal activity of isoflavonoids in different reduced stages on R.solani and S.rolfsii. Phytochemistry, 298018030031-9422
160. M. Weidenbörner, C. H. Jha, 1997Antifungal spectrum of flavone and flavanone against 34 different fungi. Mycological Research, 1017337360953-7562
161. Y. Yamada, K. Azuma, 1997Evaluation of the in vitro antifungal activity of allicin. Antimicrobial Agents and Chemotherapy, 17437490066-4804
162. T. Yoshihara, Y. Hagihara, T. Nagaoka, S. Chiba, S. Sakamura, 1988Fungitoxic compounds from the root of the eggplant stock. Annals of the Phytopathological Society of Japan, 544534590031-9473
163. J. K. Zjawiony, 2004Biologically active compounds from Aphyllophorales (Polypore) fungi. Journal of Natural Products, 673003100163-3864
164. P. H. A. Zollo, L. Biyiti, C. Tchoumbougnang, C. Menut, G. Lamaty, P. Bouchet, 1998Aromatic plants of Tropical Central Africa. Part XXXII. Chemical composition and antifungal activity of thirteen essential oils from aromatic plants of Cameroon. Flavour and Fragrance Journal, 1321071141099-1026

Chapter 3

Phytotoxicity: An Overview of the Physiological Responses of Plants Exposed to Fungicides

Maria Celeste Dias

Department of Biology & Center for Environmental and Marine Studies (CESAM), University of Aveiro, 3810 Aveiro, Portugal

ABSTRACT

In the last decades, the use of fungicides in agriculture for fungi diseases control has become crucial. Fungicide research has produced a diverse range of products with novel modes of action. However, the extensive use of these compounds in the agriculture system raises public concern because of the harmful potential of such substances in the environment and human health. Moreover, the phytotoxic effects of some fungicides are already recognized but little is known about the impact of these compounds on the photosynthetic apparatus. This paper presents a comprehensive overview of the literature considering different classes of fungicides and their effects on plant physiology, with particular emphasis on photosynthesis.

INTRODUCTION

Fungicides are chemical compounds or biological organisms that destroy or inhibit the growth of fungi or fungal spores [1]. The use of fungicides for an effective control of plant diseases has become crucial in the last decades in the agriculture system since it is estimated that fungal infections cause yield reductions of almost 20% of crops worldwide [1]. Due to their relatively low cost, ease of use, and effectiveness, fungicides became the primary means of fungi control [2]. However, the extensive use of these compounds to control fungal disease in plants raises the appearance of new strains of pathogens that have become resistant to the available commercial products [3].

Fungicide toxicity is not always restricted to the target pest organism, having also been demonstrated in mammals [4] including humans [5]. The extensive use of fungicides in plant protection against fungal disease generates long-term residues in food and in the environment [6, 7]. In the annual EU report, EFSA (European Food Safety Authority), where vegetables and fruits of 27 countries were surveyed for pesticides contamination, the results highlighted that dithiocarbamates are among the most common residual contaminants. Thus, the abusive use of such compounds in agriculture has mobilized public concern because of the harmful potential of such substances in the environment and in the food chain representing a risk for human health [6].

Most of the work dealing with the impact of fungicides in agriculture is focused on their efficiency against fungal pathogens or their residues in crops [6, 8]. Several reports appoint that some fungicides may enhance plant defences through phytoalexin synthesis and cell wall lignification or stimulate enzymes involved in the synthesis of phenolic compounds [8–10]. Others describe the putative protective role of fungicides for crops against various types of stress. Wu and Von Tiedemann [11] reported that triazoles protect Hordeum vulgareand Arachis hypogaea against ozone exposure or salt stress by stimulating antioxidative enzymes. Moreover, azoxystrobin and epoxiconazole fungicides induced a delay of senescence of Triticum aestivum mainly due to an enhancement of the antioxidative potential protecting the plants from harmful active oxygen species [11]. Muthukumarasamy and Panneerselvam [12] described an induction of the synthesis of photosynthetic pigments and proteins in fungicide-exposed plants.

However, few studies have addressed the question of whether these products alter or inhibit physiological and metabolic activities in the plant [3], and the negative effects of some fungicides on photosynthesis, pigment content, growth, and alterations in the reproductive organs were poorly explored [7, 8]. The available data report modifications on the CO_2 assimilation and photosynthetic efficiency [7, 8, 13–15]. Photosynthesis reduction strongly conditions biomass production and growth rates, which are strictly related with crop productivity and yield. Information on fungicides effects on plant physiology (photosynthesis) is crucial for the understanding of the underlying regulatory mechanisms as a precondition to judge the phytotoxicity of a compound. In this paper I present a comprehensive overview of the literature considering different classes of fungicides and their effects on plant physiology, with particular emphasis on photosynthesis. Finally, some general conclusions are drawn.

TYPES OF FUNGICIDES

Fungicides that are used to control plant fungi can be applied before infection to protect the plant from fungi invasion. This type of fungicides have a protective action. Others can be used to eliminate or eradicate an established infection. Fungicides can be classified in two main categories: contact (nonsystemic) and systemic fungicides.

Contact Fungicides

Contact fungicides have preventive action by killing or inhibiting fungi or fungal spores before the mycelia can grow and develop within the plant tissues [16]. However, once the infection is established, this fungicide may not have any function. Thus, this kind of fungicides can be used only as protectants.

Inorganic copper compounds such as Bordeaux mixture and copper carbonate and inorganic sulfur in the form of elemental sulfur and lime sulfur are some examples of the main contact fungicides available for plant protection [3]. Within the organic contact fungicides, for example, dialkyldithiocarbamates, which include the fungicides thiram, ferbam, and ziram, are a group of fungicides with a high role in the worldwide control of plant diseases since they are generally more effective and less toxic than the inorganic compounds (e.g., sulfur and copper fungicides) [4, 17]. These multisite inhibitors have several kinds of toxic action in fungal cells such as metal chelation, mixed disulfide formation, and transport of heavy metals across membranes. Dialkyldithiocarbamates inhibit a wide range of fungal enzymes, but the pyruvic dehydrogenase system is particularly sensitive to these fungicides [4]. Another group of organic contact fungicide widely used is the ethylenebisdithiocarbamates, which include zineb, maneb, metiram, and mancozeb. The mode of action of this type of fungicide differs from that of the dialkyldithiocarbamates: they undergo transformation to ethylenediisothiocyanate, which inactivates thiol groups of enzymes and metabolites in fungal cells.

Contact fungicides are inexpensive and fungal resistance rarely occurs. Therefore, they are still widely used for plant disease control even though many newer, more potent systemic fungicides have been developed

Systemic Fungicides

The other category of fungicides, systemic fungicides, are absorbed by the plant and carried by translocation to the site of infection. These kind of fungicides can kill the fungus after the mycelia has penetrated the parenchyma of the plant tissue, stopping the dispersal or infection within the plant [16]. Systemic fungicides can be used as protectants, eradicants, or both and are the most recently developed and the most promising type of fungicide for the future [3]. However, since systemic fungicides usually have a very specific site of action in the target fungus, fungi may readily develop resistance to them if they are not managed appropriately.

Systemic fungicides comprise a wide group of compounds with several modes of action. For example, the largest and most important group of systemic fungicides used to control plant fungal diseases is the dicarboximide. The mode of action of this fungicide seems to be related to the inhibition of triglyceride biosynthesis in the fungi [18]. The dicarboximide fungicides, iprodione, procymidone, vinclozolin, chlozolinate, and metomeclan are particularly valuable for the control of plant diseases caused by species ofBotrytis, Sclerotinia, Monilinia, Alternaria, Sclerotium, and Phoma.

Benzimidazoles are a group of organic fungicides with systemic action that are also extensively used in agriculture. These types of compounds control a broad range of fungi at relatively low application rates [19]. For example, benomyl is one of the most effective and extensively used benzimidazoles in crop protection [3]. The benzimidazoles benomyl, carbendazim, and thiabendazole and the phenylcarbamate diethofencarb specifically interfere with the formation of microtubules, which function in a variety of cellular processes, including mitosis and maintenance of cell shape. These fungicides bind specifically to protein subunits called tubulin and prevent their assembly to form microtubules [20].

Since their introduction in the 1960s, systemic fungicides have gradually replaced the older nonsystemic products, establishing higher levels of disease control and developing new fungicide markets [1]. Compared with the nonsystemic, systemic fungicides are approximately twice as valuable in terms of sales [21].

PLANT PHYSIOLOGICAL RESPONSES TO FUNGICIDE EXPOSURE

The widely accepted assumption that fungicide has low phytotoxicity has started to be outdated with the publication of more detailed analysis at the cell level that demonstrated several damages at the photosynthetic apparatus (e.g., [7, 8, 15]). Some reports appointed that application of fungicides has consequences on plant physiology, such as grow reduction, perturbation of reproductive organ development, alteration of nitrogen, and/or carbon metabolism [8]. This former physiological trait is fundamental for plant culture and is reflected by both photosynthetic rate and mobilization of carbohydrate reserves [7]. Physiological studies after fungicide application on several species reported modifications of both photosynthetic activity and chlorophyll a fluorescence [2, 8, 15, 22, 23]. Decreased CO_2 assimilation in fungicide-treated plants has been attributed to both stomatal (due to stomatal closure) [2] and nonstomatal effects due to a disruption in the capacity of RuBisCO carboxylation, decrease of RuBisCO content, and/or reduction of the ribulose 1.5 bisphosphate regeneration [7, 24].

Net CO_2 assimilation reductions accompanied by changes in stomatal conductance and intercellular CO_2 concentration were reported in Malus domestica and Cucumis sativus after fungicide application [2, 15]. The application of a nonsystemic fungicide, fludioxonil, in Vitis vinifera induced a decrease in net CO_2 assimilation and in the intercellular CO_2 concentration but stomatal conductance was not affected [7]. In another study reported by Saladin et al. [8], the application of the same fungicide, fludioxonil, and a systemic fungicide, pyrimethanil, in in vitro plants and fruiting cuttings of Vitis vinifera promoted different physiological responses: in in vitro plants, both fungicides decreased net CO_2 assimilation, transpiration rate, stomatal conductance, and intercellular CO_2 concentration; in the fruiting cuttings, the fungicides did not affect CO_2 exchange neither transpiration rates.

Some reports suggested that the systemic fungicide strobilurin may improve the water status and stress management of plants under conditions of drought stress [25, 26]. Nason et al. [24] showed that the application of beta-methoxyacrylate, a strobilurin fungicide, improve the water use efficiency only in well-watered Triticum aestivum and Hordeum vulgare plants. However, when these plants are under drought stress, strobilurin strongly reduces net CO_2 assimilation, intercellular CO_2 concentration, transpiration rate, and rate of stomatal conductance to water. In this study, net CO_2 assimilation reduction seems to be related to stomatal conductance decrease. It is possible that stomata respond to strobilurin-induced changes in mesophyll photosynthesis either by sensing changes in the intercellular CO_2 concentration or by responding to the pool size of an unidentified C-fixing substrate. It is also possible that the effects of strobilurin fungicides are mediated via ABA-based chemical signaling [24].

The analysis of several chlorophyll a fluorescence parameters of plants treated with fungicides [2, 15, 22, 23, 27] demonstrated that light reactions of photosynthesis are also sensible to fungicide exposure. Bader and Abdel-Basset [27] showed, for the first time, that fungicides of the triforine type (a systemic and contact fungicide) strongly inhibit electron-transport reactions of chloroplasts. Moreover, the application of systemic fungicides, benzimidazoles and triazole, and a dithiocarbamate contact fungicide affected the effective quantum yield of PSII (Φ_{PSII}) as well as the maximal quantum efficiency of PSII (F_v/F_m). This reduction was attributed to the decrease in photochemical quenching (qP) [2, 15, 23]. In Glycine max, strobilurin fungicides application reduced the ratio of F_v/F_m [24]. Strobilurin fungicides seem to block the transport of electrons between PSII and PSI by binding to the Qi site of the chloroplast cytochrome bf complex [24].

Since plants depend on photosynthesis to assimilate carbon for further growth and overall vigor, photosynthesis impairment has negative consequences in plant biomass production and yield. Several reports support a decrease in biomass production in fungicides-treated plants: benomyl, a systemic fungicide, reduced the growth of Gossypium hirsutum, Helianthus annuus, Cucumis sativus, Lactuca sativa, and Pinus taeda [3,13]. Moreover, the application of carbendazim (systemic benzimidazole fungicide) in Nicotiana tabacumaffected negatively plant biomass [3].

Pigment biosynthesis is appointed by Ahmed et al. [13] to be inhibited by the systemic fungicide, benomyl. This fungicide induces a considerable reduction on the chlorophyll a, chlorophyll b, carotenoids, and the total pigments content of Helianthus annuus plants [13]. Similarly, the treatment ofVitis vinifera with fludioxonil and Nicotiana tabacum with carbendazim also decreases the chlorophyll and carotenoid content [3, 8]. Mihuta-Grimm et al. [28] and Van Iersel and Bugbee [23] reported leaf chlorosis after benomyl application onImpatiens walleriana, Cucumis sativus, Celosia plumose, Petunia hybrid, and Lycopersicon esculentum.

Modifications of dark respiration were reported after mancozeb (contact fungicide) and flusilazol (systemic fungicide) application in Malus domestica [15]. The increase in dark respiration can be explained by additional energy requirement, metabolic breakdown of the compound, and/or activation of the alternative, cyanide-insensitive, respiration. Curiously, the treatment with strobilurin fungicides induced different responses: while in Triticum aestivum and in Spinacia oleracea plants respiration was inhibited [25, 29], in Triticum aestivumdark respiration was reduced [24].

CONCLUSIONS

The available studies in the literature have demonstrated that fungicide application impairs photosynthesis. However, the reports available are in general based on few physiological parameters using a large variety of plant species and different types of fungicide leading in some cases to controversial results that jeopardize a comprehensive knowledge of the main photosynthetic targets of fungicides. Thus, future investigation on the subject should be considered in order to produce more reliable data to identify fungicide photosynthetic targets and build a comprehensive model of the physiological response of plant exposed to fungicides.

It is expected that fungicides remain an essential tool for plant disease management and will continue to play a crucial role in optimizing yields from the world crops. Therefore, the development of new compounds with lower negative impact in plant physiology is a future challenge. This will provide benefits not only for plants yield but also for the environment and human health.

ACKNOWLEDGMENT

This work was supported by the Portuguese Foundation for Science and Technology (FCT) through a postdoctoral fellowship to M. C. Dias (SFRH/BPD/41700/2007).

REFERENCES

1. M. L. Gullino, P. Leroux, and C. M. Smith, "Uses and challenges of novel compounds for plant disease control," Crop Protection, vol. 19, no. 1, pp. 1–11, 2000.
2. X. J. Xia, Y. Y. Huang, L. Wang et al., "Pesticides-induced depression of photosynthesis was alleviated by 24-epibrassinolide pretreatment in Cucumis sativus L.," Pesticide Biochemistry and Physiology, vol. 86, no. 1, pp. 42–48, 2006.
3. P. C. García, R. M. Rivero, J. M. Ruiz, and L. Romero, "The role of fungicides in the physiology of higher plants: implications for defense responses," Botanical Review, vol. 69, no. 2, pp. 162–172, 2003.
4. F. Belpoggi, M. Soffritti, M. Guarino, L. Lambertini, D. Cevolani, and C. Maltoni, "Results of long-term experimental studies on the carcinogenicity of ethylenebis- dithiocarbamate (mancozeb)

in rats," Annual New York Academic Science, vol. 982, pp. 123–136, 2002.

5. C. A. C. Mendes, G. E. Mendes, J. P. Cipullo, and E. A. Burdmann, "Acute intoxication due to ingestion of vegetables contaminated with aldicarb," Clinical Toxicology, vol. 43, no. 2, pp. 117–118, 2005.
6. European Food Safety Authority, "Annual report on pesticide residues," EFSA Scientific Report vol. 305, 2009.
7. A. N. Petit, F. Fontaine, C. Clément, and N. Vaillant-Gaveau, "Photosynthesis limitations of grapevine after treatment with the fungicide fludioxonil," Journal of Agricultural and Food Chemistry, vol. 56, no. 15, pp. 6761–6767, 2008.
8. G. Saladin, C. Magné, and C. Clément, "Effects of fludioxonil and pyrimethanil, two fungicides used against Botrytis cinerea, on carbohydrate physiology in Vitis vinifera L.," Pest Management Science, vol. 59, no. 10, pp. 1083–1092, 2003.
9. J. P. Bryant, F. S. Chapin, and D. R. Klein, "Carbon/nutrient balance of boreal plants in relation to vertebrate herbivory," Oikos, vol. 40, no. 3, pp. 357–368, 1983.
10. C. S. Awmack and S. R. Leather, "Host plant quality and fecundity in herbivorous insects," Annual Review of Entomology, vol. 47, pp. 817–844, 2002.
11. Y. X. Wu and A. Von Tiedemann, "Impact of fungicides on active oxygen species and antioxidant enzymes in spring barley (Hordeum vulgare L.) exposed to ozone," Environmental Pollution, vol. 116, no. 1, pp. 37–47, 2002.
12. M. Muthukumarasamy and R. Panneerselvam, "Amelioration of NaCl stress by triadimefon in peanut seedlings," Plant Growth Regulation, vol. 22, no. 3, pp. 157–162, 1997.
13. A. M. Ahmed, M. D. Heikal, and O. S. Hindawy, "Side effects of benomyl (Fungicide) treatments on sunflower, cotton and cowpea plants," Phyton, vol. 23, pp. 185–195, 1983.
14. R. Untiedt and M. Blanke, "Effects of fruit thinning agents on apple tree canopy photosynthesis and dark respiration," Plant Growth Regulation, vol. 35, no. 1, pp. 1–9, 2001.
15. R. Untiedt and M. M. Blanke, "Effects of fungicide and insecticide mixtures on apple tree canopy photosynthesis, dark respiration

and carbon economy," Crop Protection, vol. 23, no. 10, pp. 1001–1006, 2004.

16. M. P. Yuste and J. Gostinear, Handbook of Agriculture, Marcel Dekker, New York, NY, USA, 1999.
17. M. Hunsche, L. Damerow, M. Schmitz-Eiberger, and G. Noga, "Mancozeb wash-off from apple seedlings by simulated rainfall as affected by drying time of fungicide deposit and rain characteristics," Crop Protection, vol. 26, no. 5, pp. 768–774, 2007.
18. L. G. Copping, "Review of major agrochemical classes and uses," in Chemistry and Technology of Agrochemical Foundations, D. A. Knowles, Ed., p. 29, Kluwer Academic Publishers, Dordrecht, The Netherlands, 1998.
19. C. J. Delp, "Benzimidazole and related fungicides," in Modem Selective Fungicides: Properties, Applications, Mechanisms of Action, H. Lyr, Ed., pp. 233–244, Wiley, New York, NY, USA, 1987.
20. C. Tomlin, "The pesticide manual: a world compendium," in Incorporating the Agrochemicals Handbook, Royal Society of Chemistry, England, UK, 1994.
21. H. G. Hewitt, Fungicides in Crop Protection, CAB International, New York, NY, USA, 1998.
22. B. W. Krugh and D. Miles, "Monitoring the effects of five "nonherbicidal" pesticide chemicals on terrestrial plants using chlorophyll fluorescence," Environmental Toxicology and Chemistry, vol. 15, no. 4, pp. 495–500, 1996.
23. M. W. Van Iersel and B. Bugbee, "Phytotoxic effects of benzimidazole fungicides on bedding plants,"Journal of the American Society for Horticultural Science, vol. 121, no. 6, pp. 1095–1102, 1996.
24. M. A. Nason, J. Farrar, and D. Bartlett, "Strobilurin fungicides induce changes in photosynthetic gas exchange that do not improve water use efficiency of plants grown under conditions of water stress," Pest Management Science, vol. 63, no. 12, pp. 1191–1200, 2007.
25. J. Glaab and W. M. Kaiser, "Increased nitrate reductase activity in leaf tissue after application of the fungicide Kresoxim-methyl,"

Planta, vol. 207, no. 3, pp. 442–448, 1999.

26. K. Grossmann, J. Kwiatkowski, and G. Caspar, "Regulation of phytohormone levels, leaf senescence and transpiration by the strobilurin kresoxim-methyl in wheat (Triticum aestivum)," Journal of Plant Physiology, vol. 154, no. 5-6, pp. 805–808, 1999.
27. K. P. Bader and R. Abdel-Basset, "Adaptation of plants to anthropogenic and environmental stresses," inThe Effects of Air Constituents and Plant-Protective Chemicals, M. Pessarakli, Ed., Handbook of Plant and Crop Stress, pp. 973–1010, Marcel Dekker, New York, NY, USA, 1999.
28. L. Mihuta-Grimm, W. A. Erb, and R. C. Rowe, "Fusarium crown and root rot of tomato in greenhouse rock wool systems: sources of inoculum and disease management with benomyl," Plant Disease, vol. 74, pp. 996–1002, 1990.
29. H. Köhle, K. Grossmann, G. Retzlaff, and A. Akers, "Physiological effects of the new fungicide Juwel on yield in cereals," Gesunde Pflanzen, vol. 49, no. 8, pp. 267–271, 1997.

Chapter 4

Hormesis: Biphasic Dose-Responses to Fungicides in Plant Pathogens and Their Potential Threat to Agriculture

Carla D. Garzon and Francisco J. Flores

Department of Entomology and Plant Pathology, Oklahoma State University, Stillwater, OK, USA

INTRODUCTION

Hormesis is a toxicological concept characterized by low-dose stimulation and high-dose inhibition (1, 2, 3, 4, 5). Extensive examinations of scientific literature by Calabrese and his collaborators reported that hormetic dose-responses are common across biological systems and stressors (4, 6, 7, 8). Scientific literature provides evidence that hormesis can be caused by multiple stimuli (9), such as chemicals (4, 10, 11), radiation (12; 29), heat (13), stress (14), and even exercise

(15). Dose-response curves displaying hormesis are characterized by a biphasic behavior (Fig.1). The hormetic zone includes a range of subinhibitory doses that are stimulant, with a peak at the maximum stimulation dose (MSD), and ends at the no observable adverse level (NOAEL), that typically precedes the inhibitory doses (Fig. 1). Our interest in hormesis pertains to the effects of fungicides at subinhibitory doses on fungal and oomycete growth and pathogenicity. Thereafter, for the purposes of this book, we will focus on chemical hormesis alone. Some of the most familiar examples of biphasic dose-responses include vitamins, alcohol, essential minerals, and many drugs (16, 17, 18, 19, 20). Hormesis has been measured using diverse endpoints in multiple biological systems (8). One of the most common endpoints used in hormesis research is growth, but several others have been studied including CO_2production (1; 21), longevity (14), other metabolic processes, and cellular functions (22). Southam and Ehrlich (2) coined the term hormesis to describe the biphasic dose-response because its Greek etymological root *horm-* means "to exite" (18). Several reports of subinhibitory stimulation of fungi and oomycetes due to exposure to fungicides are available in the mycological and phytopathological literature, but little attention has been paid to fungicide hormesis in spite of its potential detrimental effects to crop productivity.

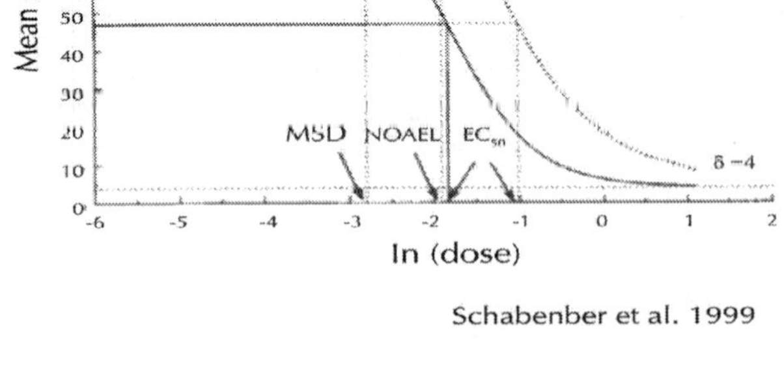

Figure 1: Dose-response curves comparing the traditional threshold model (solid) and the Brain-Cousens model (dotted) that includes hormesis in the dose-response concept. The Brain-Cousens model permits to estimate the

no-observable adverse effect level (NOAEL), the maximum stimulation dose (MSD), and the half maximal effective concentration (EC_{50}). Modified from Shabenberg *et al.* (23).

HISTORIC BACKGROUND

The first references to dose-responses have been attributed to Paracelsus, who allegedly said "all things are poison and nothing is without poison, only the dose permits something not to be poisonous" (24). However, the first scientific reports of chemical stimulation at low doses and inhibition at high concentrations go back to the 1800s. In 1854, Virchow (22) reported that low concentrations of sodium hydroxide increased the frequency and intensity of beating of ciliae of human tracheal ciliated epithelia; while at higher concentrations or longer exposures the same compound paralyzed ciliae and caused cell death. In 1865, Reveil (25) reported that sodium hypochlorite stimulated seed germination at low concentrations (0.1% solution) but it was phytotoxic at high concentrations. According to Calabrese and Baldwin, in their review of the historical foundations of chemical hormesis as a biological hypothesis (5), two researchers are considered as the founders of modern hormesis, Rudolf Arndt and Hugo Schulz. In 1887, Schulz (1) observed experimentally that several chemicals caused low-dose stimulation and high-dose stimulation on yeast fermentation. Because Schulz's views supported those of Arndt, a homeopathic physician, soon they were merged in to what was known as the "Arndt-Schulz law". The law stated that "for every substance, small doses stimulate, moderate doses inhibit and large doses kill" and that their findings could be generalized to all organisms and all toxic agents. Mainly because of the difficulty at the time to demonstrate the universality of this law without offering explanation of its biological causes, and probably also because at the time it was conceptually associated to homeopathy, over time it fell out of use. However, a few years later Ferdinand Hueppe (1896), a distinguished bacteriologist, made similar observations in bacteria and described the phenomenon in his books and scientific publications (26). Hueppe recognized the validity of Schulz's scientific research but stated certain limitations and exceptions to the Arndt-Schulz law. Soon the notion that "substances which inhibit biological processes at sublethal doses may be expected to stimulate them at lower levels" became known as the "Hueppe rule" and was broadly adopted in international literature (3, 5).

Years later, in 1929, Branham (21) confirmed Schulz's observations using a series of different chemicals and an improved apparatus to detect CO_2 production, and demonstrated that very small doses of inhibitor compounds had an apparent stimulatory effect on carbon dioxide production by yeasts. One of her experiments assessed the effect of adding crystals of 1,2,5,6-dibenzanthracene to yeast suspensions, finding that at a concentration of 9×10^{-4} molar yeast proliferation increased. The term hormesis was coined by Southam and Ehrlich in their 1943 report of growth stimulation of a wood-decaying fungus (*Fomes officinalis*) in culture by extracts of western red-cedar heartwood at low doses, while higher doses were inhibitory (2).

Research focused on chemical subinhibitory dose-responses continued through the early decades of the 1900s, but scientific attention faded away as the science of Toxicology became established and few hormesis studies were published until the 1980s (3, 27, 28). Toxicology is the study of the adverse effects of chemical, physical, or biological agents on biological systems, their prevention and amelioration. Hence, by definition, Toxicology deals with the negative effects of such agents at doses above the no observed adverse effect level (NOAEL), and essentially ignores the effects of subinhibitory doses. Stebbing (3) reexamined the hormesis concept and provided the first update on the validity of this concept based on the abundance of scientific reports of data with biphasic distributions. Stebbing reported his conclusions after searching the scientific literature for an explanation to his own observations of growth stimulation of the colonial hydroid *Campanularia flexuosa* (3) by exposure to sublethal concentrations of various metals and organometallic compounds; and raised the question whether growth hormesis by diverse toxic substances in multiple biological systems had a common explanation. Although examples of stimulation by subinhibitory levels of chemicals and radiation are numerous in the toxicological literature, interest on hormesis has slowly increased again in the last few decades. Several studies on radiation hormesis followed Stebbing's paper (29, 30, 31, 32), with limited initial impact initially, but the interest on the subject gathered momentum by the end of the decade (33, 34). The few chemical hormesis reports that followed Stebbing's paper had much better acceptance (35, 36, 37). Hormesis once again, although still viewed as a polemic concept by some toxicologists, attracted the attention of the scientific community on the basis of its relevance to risk assessment and optimization of research resources (38). In the late 80s Dr. Edward Calabrese, currently the Director of the Department of Public Health at University of Massachusetts School of Public Health and Health Sciences, became deeply interested in hormesis and began a campaign to create awareness of the biological significance of this phenomenon.

Calabrese first report related to hormesis described stimulation of *Mentha piperita* grown, in soils and*in vitro,* by the growth retardant phosfon (2,4 dichlororobenzyl tributyl phosphonium chloride) at concentrations of $1.26x10^{-5}$ M to $7.77x10^{-4}$ M and $6.30x10^{-7}$ M to $3.78x10^{-5}$ M respectively, while higher concentrations inhibited plant growth (39); nonetheless, no reference to hormesis was made, but wondered about the nature of the stimulation. Eleven years later, Calabrese and collaborators examined the occurrence of chemically induced hormesis in biological and toxicological systems by looking for evidence of low-dose stimulation in literature (37). They found multiple historical and contemporary publications reporting biphasic dose-responses to chemical stressors in plants, fungi, and animals. During the decade that followed Calabrese and Baldwin published four articles reviewing possible examples of chemical hormesis in previously published studies, and discussing the importance of expanding the reference dose concept to incorporate the effect of subinhibitory doses (4, 40, 41, 42). Products of their work were the description of the first quantitatively-based methodology to evaluate chemical hormesis and the development of a chemical hormesis database. These publications served as catalysts for what Calabrese and Baldwin referred to as "the dose-response revolution". The prolific research inspired by these papers has produced hundreds of publications, providing well supported evidence of hormetic effects of a broad spectrum of stressors, including radiation, heat, caloric intake, and even exercise, on plants, fungi, bacteria, protozoa, and animals, including humans (6, 10, 12, 43, 44, 45, 46, 47).

HORMESIS AS A GENERAL PHENOMENON

Based on their extensive literature review, Calabrese and Baldwin stated that hormetic responses often, but not always, display the following characteristics: i) Stimulation zone of the dose-response could be found within a 10-fold range; ii) Stimulatory responses were 30-60% greater than the controls; and iii) the NOAEL three to six-fold greater than the MSD (4). Using these criteria they identified hundreds of toxicological studies that potentially displayed hormetic responses (48). In recent years researchers from many different disciplines have been inspired by Dr. Calabrese's work and have studied hormesis in their biological systems of interest. There are numerous recent reports of biphasic dose-responses in plants, animals, as well as eukaryotic and prokaryotic microorganisms in the scientific literature. A few examples of studies reporting chemical hormesis are presented below.

Velini *et al.* (49) examined the effect of the herbicide Glyphosate on target and non-target plants. The growth of glyphosate sensitive soybean was inhibited when applied at concentrations between 72–720 g AE ha^{-1}; however, low doses (1.8 – 18 g AE ha^{-1}) of herbicide induced significant increases in shoot and total dry weight (up to 28% and 22%, respectively). Similar growth enhancements were observed in maize, *Eucalyptus grandis* Hill ex Maiden, *Pinus caribea* L. and *Commelina benghalensis* L. by 1.8 – 36 g AE ha^{-1}. Barceló and Poschenrieder (10) reported their observations of rapid root growth in corn plants sensitive to soil aluminum (Al). Hormesis was observed in plants exposed to subinhibitory Al levels or as a transient effect after brief exposure to potentially toxic concentrations. A study by Migliore *et al.* (50) of the effects on plants of fluoroquinolone antibiotics administered to cattle and excreted in their feces, which are later used as field manure, demonstrated that enrofloxacin was toxic to *Cucumis sativurs* L., *Lactuca sativa* L., *Phaseolus vulgaris* L., and *Raphanus sativus* L. seedlings at concentrations equal or above 5000 μg l^{-1}, while growth hormesis was often observed at concentrations 50-100 μg l^{-1} in the four plant species.

The better documented example of hormesis in animals is lifespan increase as a result of restricted caloric intake in diet (24). While high calorie diets have been associated with increased risk of several age-related diseases in animal systems (cardiovascular disease, type 2 diabetes, stroke, among others), dietary energy restriction (i.e. controlled caloric restriction or intermittent fasting) has been reported to have anti-oxidative effects, increasing the cells' tolerance to several types of stresses. For example, restricted calorie diets protected rodents against several types of cancers (51); furthermore, alternate day calorie restricted diet in humans seems to improve inflammatory symptoms in asthmatic patients (52). A now classic example of chemical hormesis are vitamins in human diet, since small amounts of them are necessary and beneficial, but large amounts are toxic and can cause hypervitaminosis, tissue mineralization, and chemical imbalances (45). Other examples of hormesis in animal systems include inhibition of N-diethylnitrosamine (DEN)- initiated pre-neoplastic lesions by phenobarbital at low-doses, while higher doses promote activity (53), and survival and fertility enhancement in *Podisus distinctus* Stål (Heteroptera: Pentatomidae) due to exposure to sublethal doses of the pyrethroid insecticide permethrin (54), among others.

There are abundant examples of hormesis in prokaryotic and eukaryotic microbial systems. As related in the historical review, the first reports of biphasic dose-responses were on bacteria and fungi (1, 2, 26, 21) and several more studies have been published in more recent years. Hotchkiss (55) found that $TiCl_2$, $MgCl_2$ and, NaCl had hormetic effects on the growth of *Escherichia coli* in culture. Low doses of penicillin produced doubled the growth of *Staphylococcus* (No. 6571 N.C.T.C.) in culture compared to the non-treated control (56). Linares *et al.* (57) demonstrated that three antibiotics (tobramycin, tetracycline, and norfloxacin) trigger the expression of determinants that influence the virulence of *Pseudomonas aeruginosa* at subinhibitory concentrations. Wang *et al.* (58) and Gong *et al.* (59) reported increased production of mycrocystin by *Microcystis aeruginosa* (cyanobacteria, Cyanophyta) in non-linear responses to nonylphenol and arsenic pollution. Hong et al (60) reported growth stimulation in *Selenastrum capricornutum* (Chlorophyta, Selenastraceae) due to exposure to subinhibitory doses of the algicide ethyl 2-methyl acetoacetate at high initial algal densities. Many of the early studies on hormesis were done on the effects of multiple chemicals on yeast metabolism (2, 21). Yeasts continue to be used as models for the study of hormesis, particularly related to cancer (61), ageing (43, 62), and UV radiation hormesis research (63).

EVIDENCE OF CHEMICAL HORMESIS IN PHYTOPATHOLOGICAL LITERATURE

Our review of mycological and phytopathological literature found several studies of fungicide effects on fungi and oomycetes with results that reflect potential hormetic responses. We present some interesting examples below, while an exhaustive literature review will be reported elsewhere. Southam and Ehrlich (2) observed that extracts of western red-cedar heartwood were stimulatory at low doses on the growth of a wood-decaying fungus (*Fomes officinalis*) in culture, while higher doses were inhibitory. This was the first scientific study that demonstrated stimulation by a mycotoxic compound on a plant pathogenic fungus, and the authors coined the term hormesis to describe their observations. Later studies of hormesis in plant pathogens demonstrated a positive effect of trichothecin, a compound produced by *Trichothecium roseum,* on the growth of *Fusarium oxysporum* (64, 65) while assessing the production of trichothecin in different soil types. Although not further references to hormesis were made in the phytopathological literature until recently (66), several reports of stimulation by exposure to fungicide are available. Baraldi *et al.* (67) reported higher percentages of germination in seven out of 41 thiabendazole (TBZ) resistant isolates of *Penicillium expansum* from pear when grown TBZ-amended media than without the fungicide. The authors suggested that germination stimulation and fitness advantage in certain TBZ-resistant *P. expansum* isolates could be due to the ability of this isolates to metabolize the fungicide as a nutrient compound, but further study was recommended to understand the nature of the stimulation. Audenaert *et al.* (68) observed increased production of the mycotoxin deoxynivalenol (DON) by *Fusarium graminearum, in vitro* and *in planta,* when exposed to sub-lethal doses of the triazole fungicide prothioconazole, and proposed that mycotoxin production was stimulated by H_2O_2 production triggered by the fungicide at low concentrations. While studying the effects of different fungicides on the infection of *Sphagnum* by fungi, Landry *et al.* (69) observed significantly increased radial growth of *Lyophyllum palustre* (Peck) *in vitro* when media was amended with the fungicide propamocarb compared to the non-amended control.

Similar references can be found on oomycetes literature. Fenn and Coffey (70) observed that 69 μg/ml of phosphorous acid (H_3PO_3) was stimulatory on the growth of *Pythium ultimum in vitro* and that 138 μg /ml was also stimulatory on the growth of *Pythium myriotylum*. In a study comparing the sensitivities of various oomycetes to the fungicides mefenoxam (a metalaxyl enantiomer, FRAC code: 4) and hymexazole (FRAC code: 32), Kato *et al.* (71) observed a range of responses to these fungicides among the different groups that could be used to classify them by taxa as reflected by DNA analysis. Significant findings included the differential fungicide sensitivity of the plant pathogens*Pythium* and *Phytophthora*. Although fungicide sensitivity responses varied within the two genera, general trends suggested that *Pythium* species were more sensitive to hymexazol than to mefenoxam, while the opposite was true for *Phytophthora* species, with a few exceptions. The reported sensitivity response curve for *Phytopthora undulata* reflected slight stimulation at the lowest hymexazol doses. Radial growth stimulation was observed in three out of four metalaxyl-resistant *Phytophthora infestans*isolates when grown on cleared lima bean agar medium amended with 20μl/ml (72). Since one of the isolates grew more in the presence of metalaxyl only when nutrients were limited, it was hypothesized that under certain circumstances metalaxyl could be beneficial to metalaxyl-resistant *P. infestans*strains. Moorman and Kim (73) reported for the first time strains with dual resistance to mefenoxam and propamocarb in *Pythium aphanidermatum, P. irregulare,* and *P. ultimum* and described radial growth stimulation in some strains of the three species by propamocarb at a concentration of 1μg/ml, and of *P. aphanidermatum* by 1,000 μg/ml propamocarb.

FUNGICIDES HORMESIS AND ITS IMPACT ON FUNGAL PLANT PATHOGENS

Recent research on chemical hormesis on fungal pathogens has focused on the effects of subinhibitory doses of fungicides on radial growth and pathogenicity of fungi and oomycetes (66, 74). Garzon *et al.*(66) examined the effects of subinhibitory doses of mefenoxam on the radial growth and pathogenicity of a mefenoxam and propamocarb-resistant isolate of *P. aphanidermatum*. In this study we found modest radial growth stimulation, with an average of 10% increase over the control, and very significant increase of pathogenicity, with an increase of 61% severity of damping-off of seedlings in geranium. Flores and Garzon (74) reported standardized laboratory and statistical protocols for detection of chemical hormesis using radial growth *in vitro* as endpoint. Using the reported methods hormetic responses were detected in *Pythium aphanidermatum*, *Rhizoctonia solani* and *R. zeae* exposed to ethanol (Fig. 2), and on *P. aphanidermatum* exposed to subinhibitory doses of the fungicides propamocarb and cyazofamid (Fig. 3).

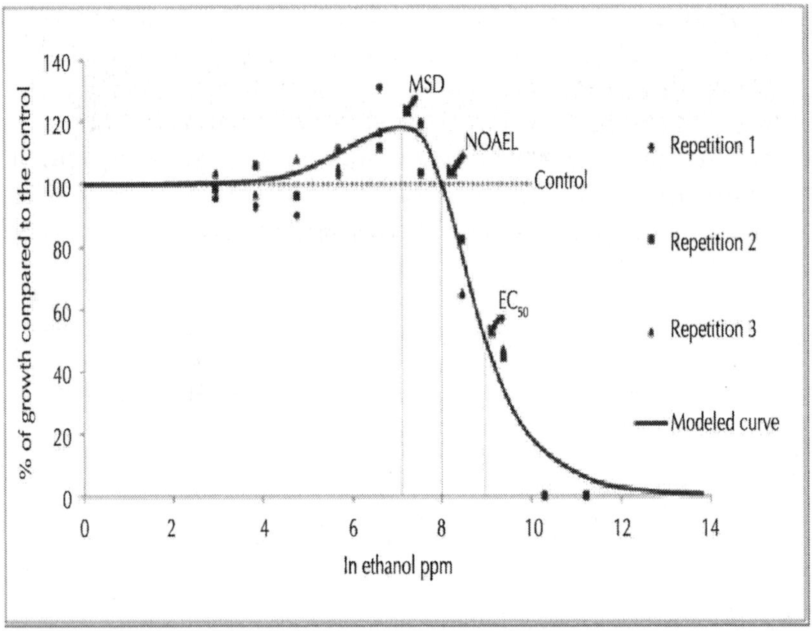

Figure 2: Modeled curve of the radial growth *in vitro* of *P. aphanidermatum* in response to subinhibitory doses of ethanol (Flores and Garzon *In press*). Radial growth is expressed as percentages relative to a non-amended control, and concentrations as natural logarithm of ppm. Figure reproduced with permission of Dose-Response Journal.

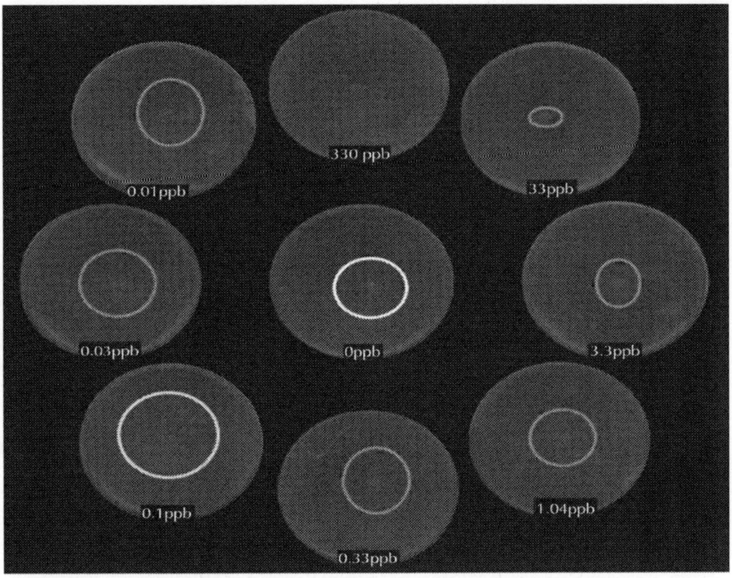

17 % radial growth increase at 0.1ppb

Figuer 3: Radial growth of *Pythium aphanidermatum* on corn meal agar amended with cyazofamid. Significant hormetic stimulation was observed at 0.1 ppb cyazofamid, with 17% radial growth increase (yellow) over the control (white).

THE BIOLOGICAL BASIS OF HORMESIS

Hormesis can result from overcompensation after a disruption of homeostasis by stressors, by direct stimulation, or as a response to an adaptating dose followed by a larger dose (3, 75, 76). The research by Branham (21) on the effect of 16 chemicals on CO2 production by Baker's yeast, provided clear evidence, in 12 of the chemicals, of an initial mild inhibition followed by significant stimulation. These results supported the hypothesis of stimulation due to overcompensation, being most evident for formaldehyde, phenol, iodine, and metaphen. Other examples of over-compensatory responses include ethanol stimulation of locomotion in mice (77), increased serotonin levels in rat neurons after treatment with below toxic doses of 5,6-dihydroxytryptamine (78); and growth stimulation in peppermint following an initial decrease after treatment with phosfon, a plant growth regulator (39), among others. Stebbing (79) provided an "improved" explanation for hormesis due to overcompensation by describing a model using two overlapping curves, an *effect* curve and a *response*curve, relative to a particular endpoint and the compensatory mechanisms involved, respectively. Under this model, below a threshold an stressor may be undetectable by an organism, and after reaching that threshold compensatory mechanisms would be triggered by a range of doses (represented by the *response* curve), however no effects would be visible, due to neutralization by compensation responses; effects would be evident only after the capacity to neutralize inhibition is exceeded and may follow a linear pattern. In systems where compensation responses completely neutralize inhibition, the alfa curve is present (threshold dose-response) with no evident low-dose effects, while in those where compensation responses exceed inhibition the beta curve is observed (hormetic curve). Overcorrection of inhibition may have an adaptation role with impact on fitness. This hypothesis was supported by Stebbing's observations that hydroid colonies pre-exposed to a subinhibitory dose of copper (10 $\mu g.l^{-1}$) had increase tolerance across the range(0 – 50 $\mu g.l^{-1}$), when compared with non-pre-exposed colonies. Calabrese suggested direct stimulatory response as another possible cause of hormesis (8).

The underlying mechanisms that generate hormetic responses have not yet been fully understood. Conolly and Lutz (80) hypothesized that hormetic responses may occur due the superimposition of two monotonic dose-responses, one that takes effect at low doses and other that overtakes at higher doses undermining the first one. They demonstrated by computational modeling that four different cellular models could generate biphasic dose-responses: i) Membrane receptor subtypes with opposite downstream effect; ii) Androgen receptor mediated gene expression; iii) Induction of DNA repair and "co-repair" of background DNA damage; and, iv) Modulation of the cell cycle and effect on rate of mutation (80). Subsequent studies have found empirical evidence of hormesis attributable to the presence of antagonistic membrane receptors (81) or to the induction of DNA repair (82). Bae *et al.*(83) suggested that hormesis may arise because of the heterogenic susceptibilities of different tissues to the same stimulus; such difference can result in the expression of a U shaped dose-response curve. The observations that drove this conclusion were made from the response of different cell types normally present in human blood vessels to the presence of small doses of arsenic and a reactive oxygen species generator (menadione). Allender (84) and Allender *et al.* (85) provided indirect evidence of the influence of calcium influx to the cell on hormetic responses related to plant growth. The diversity of the models that may show a hormetic response suggests that the mechanisms acting may not be the same for different systems. Experimental evidence suggests that multiple metabolic processes may be involved in hormetic responses, some acting during the stimulation phase and others, probably different, acting during the inhibition phase of the beta curve.

STUDYING HORMESIS

Detection of hormesis is often challenging due to the multiple factors that can affect metabolic responses of the target organisms. For example, when studying fungicide hormesis in oomycetes using radial growth as endpoint it is fundamental to standardize every experimental factor involved; in addition to growing media type and concentration, fungicide treatments, and incubation temperature, other factors are also relevant, including light, growing media depth, inoculum age and developmental stage, mixing time when preparing fungicide dilutions, using fungicide stock solutions prepared on the same day of the experiments, etc. Variation in any of these parameters can influence mycelial growth significantly, hence introducing experimental variation that could affect the reproducibility of results (66).

When trying to prove the existence of hormesis there are some requirements that the experimental design should fulfill: i) The NOAEL should be determined; ii) doses below the NOAEL need to be tested with five equally spaced doses providing enough data to detect hormesis; and iii) the separation between doses should generally be smaller than one order of magnitude since the hormetic zone is usually within a ten-fold range (42). To test for hormesis researchers must compare the effect of small doses with the response of the non-treated control. Therefore, there should always be background incidence in the control, without background incidence there is no way to detect a stimulus

(80). Evaluation of data is very important when proving hormesis. Crump suggests the criteria for evaluating hormesis as follows: strength of evidence, soundness of data, consistency and biological plausibility (86). Statistical analyses should be performed in order to differentiate a small stimulus from background occurrence.

Different methods have been used throughout the years for the detection and estimation of hormesis including parametric, non-parametric, and model-based approaches. The hormetic zone of a dose-response curve follows a non-monotonic relationship between two variables, similar to what is known as umbrella alternatives. Umbrella alternatives are important in many fields of science; a classical example is the ability of learning as a function of age in humans (87). As we grow older our ability to learn new things reaches a peak and later declines. Tests for umbrella alternatives can be used to detect if a dose-response curve follows a non-monotonic trend compared to a monotonic one where hormesis would not be present. The firsts to describe a test for umbrella alternatives were Mack and Wolfe (87) who used a non-parametric method where the distribution of the data is not assumed a-priori. In the Mack and Wolfe test (87) the maximum stimulation detected experimentally is compared to the response at all the other doses using Mann-Whitney counts, a test statistic is calculated and compared with simulated critical values to determine if the dose-response is biphasic. Buning and Kossler (88) demonstrated that the Mack and Wolfe (87) test with Mann-Whitney counts is appropriate for testing data with symmetric and medium-up to long tailed distributions but they suggested the use of different two-sample statistics, i.e. Hogg *et al.* (89) and Gastwirth (90), for asymmetric and short-tailed distributions respectively.

For the detection of hormesis there are also parametric analyses which assume a normal distribution of the data and can have more statistical power than non-parametric analyses if such assumptions are correct. Among the parametrical tests we can highlight the one proposed by Buning and Kossler (88), a modification for umbrella alternatives of the test by Barlow *et al.* (91) for monotonic alternatives. And the method by Bailer and Oris (92) that employs generalized linear models for the detection of hormesis. There are also parametric models that can be used to detect the hormetic response. A parametric model is an equation with a finite number of parameters that describe the relationship between two variables, in the case of hormesis it describes a byphasic relationship between dose and response. If the data fits the model within a confidence limit, usually of 95%, it is assumed that hormesis is present. There are some models that can be modified to better fit the hormetic response, including the quadratic function (93), Gompertz function (94, 95) and logistic function (96). Deng *et al.* (97) summarized some of these model-based approaches and developed a method to estimate the magnitude of the hormetic response. When testing for chemical hormesis the Brain and Cousens model (96), based on the logistic function, is the most commonly used. Further modifications of the Brain and Cousens model have been made by Van Ewijk and Hoekstra (98), and Schabenberger *et al.* (23) in order to allow estimation of different parameters that are relevant for toxicology such as EC_{50}, NOAEL and maximum stimulation dose (MSD). According to Cedergreen *et al.* (99), a deficiency of the Brain and Cousens model is that if the slope at the EC_{50} (β) is smaller than 1 then the model will not yield a curve; therefore they proposed a model that circumvents this drawback. Cedergreen *et al.*(99) provide a robust detection method but it does not estimate the EC_{50}, NOAEL or MSD. In general, a combination of methods may be used to detect hormesis, determine its magnitude and estimate the EC_{50}, NOAEL and MSD, depending on the type of data that is being analyzed and on the behavior of the dose-response curve.

When testing for the stimulation at low doses of chemicals on fungal plant pathogens, Flores and Garzon (74) used a model-based approach. The Brain and Cousens model was appropriate for this case where most of the dose-response datasets analyzed yielded a β higher than 1. Because of the relevance of EC_{50}, NOAEL, and MSD for disease management, the modified model by Schabenberger *et al.*(23) was used. Flores and Garzon (74) tested 9 to 11 doses of different chemicals including pesticides and disinfectants on *Pythium aphanidermatum* and *Rhizoctonia zeae* with at least five doses below the NOAEL. Datasets were analyzed using a non-linear modeling procedure (PROC-NLIN, SAS 9.2, SAS Institute, Cary, NC) and EC_{50}, NOAEL, and MSD were estimated as described by Schabenberger *et al.* (23). The procedure yielded dose-response curves that properly described the behavior of the data and showed their biphasic nature (Fig 4).

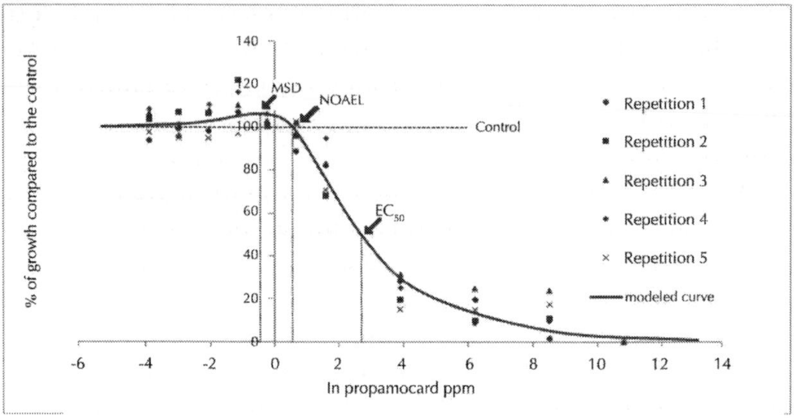

Figure 4: Observed values and modeled curve of the radial growth *in vitro* of *P. aphanidermatum* in response to low doses of propamocarb. Radial growth is expressed as percentages relative to a non-amended control, and concentrations as natural logarithm of ppm. Each data point in the figure represents the mean value of five replicates. Arrows indicate the natural logarithm of the EC_{50} (14.91 ppm), NOAEL (1.77 ppm) and MSD (0.64 ppm). The slope of the curve at the EC_{50} (β) was 1.37 and the rate of increase at low doses (?) was 35.17. Low-dose stimulation was observed (Flores and Garzon, [74]). Figure reproduction authorized by Dose-Response.

WHY IS CHEMICAL HORMESIS RELEVANT FOR CROP MANAGEMENT?

Multiple chemicals with distinct modes of action are available for management of fungal and oomycete diseases. Although integrated disease management is practiced extensively, the productivity of many agricultural systems relies strongly on chemical control. The limited access to registered products for certain agricultural environments, such as greenhouses, as well as inappropriate use has led to the emergence of fungicide resistant strains in multiple species (100, 101, 102, 103, 104, 105, 106). Currently, the effects of subinhibitory doses of fungicides on fungal plant pathogens are unknown. The evidence gathered from literature indicates that stimulation of fungi and oomycetes by subinhibitory doses of fungicides has been observed in ascomycetes (64, 65, 67, 68), basidiomycetes (2, 69, 74), as well as in oomycetes (66, 70, 71, 72, 73, 74). Several fitness factors could be affected for the benefit of pathogens, including mycelial growth, spore germination, toxin production and pathogenicity (66, 67, 68, 74). Exposure to subinhibitory doses can occur accidentally in agricultural fields, orchards, nurseries and greenhouses, under diverse circumstances, such as inappropriate fungicide application, low-dose applications to reduce costs, presence of fungicide resistant strains, etc. Thereafter, the possibility of fungal pathogen stimulation due to fungicide hormesis in actual agricultural scenarios is real. The potential effects of fungicide hormesis are highly detrimental, since it could result in larger crop losses, reduced seed and crop quality, higher mycotoxin levels in grain, and wasteful use of fungicides.

In spite of the potential detrimental effects of fungicide hormesis on fungal plant pathogens to agricultural productivity, a complete lack of awareness has meant the exclusion of this important concept from the design of disease chemical management strategies. In some fungicide-pathogen systems, the value of the EC_{50} can be different when hormesis is included in the analysis, hence it is important to consider this concept to avoid bias in EC_{50} calculations. Awareness of the risk taken by growers by the inappropriate use of reduced-dose fungicides (reduced-dose fungicides can be used in combination with two or more other formulations, with different active ingredients [107]) and careless chemical application will help to promote the use of best-management practices and responsible use of fungicides.

More research is needed to understand the processes involved in fungicide hormesis, the prevalence of hormesis in fungi and oomycete species and populations, fungicide class risks, whether mixtures can prevent stimulation, etc. Hormesis is not a new concept but its use in plant pathology is recent, and its application to disease management may open new opportunities to improve plan health and crop productivity.

REFERENCES

1. H. Schulz, 1888Ueber mefegifte. Plügers Archiv fur die gesamte Physiologic 42:517.

2. C. M. Southam, and J. Erhlich, 1943Effects of extracts of western red-cedar heartwood on certain wood-decaying fungi in culturePhytopathology 33517524
3. A. R. Stebbing, 1982Hormesis the stimulation of growth by low levels of inhibitors. The Science of the Total Environment3213234
4. E. J. Calabrese, and L. A. Baldwin, 1997bThe dose determines the stimulation (and poison): Development of a chemical hormesis database. International Journal of Toxicology16545559
5. E. J. Calabrese, and L. A. Baldwin, 1999Chemical hormesis: its historical foundations as a biological hypothesisToxicologic Pathology27195216
6. E. J. Calabrese, and L. A. Baldwin, 2001aRadiation hormesis: its historical foundations as a biological hypothesis.Human and Experimental Toxicology 194175
7. E. J. Calabrese, and L. A. Baldwin, 2001b U-s. h. a. p. e. d. Dose-responses, in Biology, Toxicology, and Public Health. Annual Review of Public Health221533
8. E. J. Calabrese, and L. A. Baldwin, 2001c. Hormesis, a. Generalizable, and unifying hypothesis. Critical Reviews in Toxicology31353424
9. E. J. Calabrese, and L. A. Baldwin, 2003a. Hormesis, the dose-response revolution. Annual Review of Pharmacology and Toxicology43175197
10. J. Barcelo, and C. Poschenrieder, 2002Fast root growth responses, root exudates, and internal detoxification as clues to the mechanisms of aluminium toxicity and resistance: a reviewEnvironmental and Experimental Botany487592
11. E. J. Calabrese, and L. A. Baldwin, 2000Chemical hormesis: its historical foundations as a biological hypothesisHuman and Experimental Toxicology 1231
12. L. E. Feinendegen, 2005Evidence for beneficial low level radiation effects and radiation hormesis.The British Journal of Radiology7837
13. S. L. Rea, D. Wu, J. R. Cypser, J. W. Vaupel, and T. E. Johnson, 2005A. Stress-sensitive, reporter predicts longevity in isogenic populations ofCaenorhabditis elegans. Nature Genetics37894898

14. J. R. Cypser, and T. E. Johnson, 2002Multiple stressors in Caenorhabditis elegans induce stress hormesis and extended longevity.Journal of Gerontology: Biological Sciences 57A: B109B114.
15. M. C. Gomez-cabrera, E. Domenec, and J. Vina, 2008Moderate exercise is an antioxidant: Upregulation of antioxidant genes by training.Free Radical Biology and Medicine 44126131
16. E. J. Calabrese, and L. A. Baldwin, 2003Ethanol and hormesis. Critical Reviews in Toxicology33407424
17. E. J. Calabrese, and L. A. Baldwin, 2003Toxicology rethinks its central belief.Nature421691692
18. A. W. Hayes, 2008Principles and Methods of ToxicologyInforma Healthcare U.S.A. Inc. New York.
19. X. K. Yu, W. M. Zhao, J. F. Ma, X. Q. Fu, and Z. Z. J. Zhao, 2011Beneficial and harmful effects of alcohol exposure on Caenorhabditis elegans wormsBiochemical and Biophysical Research Communications412757762
20. K. R. Sridhary, and F. Bärlocher, 2011Reproduction of aquatic hyphomycetes at low concentrations of Ca2, Zn2, Cu2, and Cd. Environmental Toxicology and Chemistry 3028682873
21. S. E. Branham, 1929The effects of certain chemical compounds upon the course of gas production by baker's yeast.Journal of Bacteriology4247264
22. 6133134Virchow R. 1854. Uber die Erregbarkeit der Flimmerzellen. Virch Arch 6:133-134. In: Henschler D. (2006). The origin of hormesis: historical background and driving forces. Human and Experimental Toxicology 25: 347-351
23. O. Schabenberger, B. E. Tharp, J. J. Kells, and D. Penner, 1999Statistical tests for hormesis and effective dosages in herbicide dose-responseAgronomy Journal91713721
24. M. T. Mattson, and E. J. Calabrese, 2010Hormesis: A Revolution in Biology, Toxicology and Medicine.Humana Press, Springer New York.
25. P. O. Reveil, 1865Recherches de physiologie vegetale. In: De l'action des poisons sur les plantes. Paris.
26. 27195216Hueppe F. 1896. Principles of Bacteriology. Open Court Publishing Company, Chicago. In: Calabrese EJ and Baldwin

LA (1999). Chemical Hormesis: Its Historical Foundations as a Biological Hypothesis. Toxicologic Pathology 27:195-216

27. W. A. Randall, C. W. Price, and H. Welch, 1947Demonstration of hormesis (increase in fatality rate) by penicillin.American Journal of Public Health 374421425
28. C. C. Campbell, and S. Saslaw, 1949Enhancement of growth of certain fungi by streptomycin.Proceedings Of The Society For Experimental Biology And Medicine3562562
29. H. M. Prichard, 1984Air ionization and radiation hormesis. Health Physics4611391140
30. R. J. Hickey, E. J. Bowers, and R. C. Clelland, 1983Radiation hormesis, public health, and public policy: a commentary.Health Physics44207219
31. G. S. Roessler, 1987Radiation Hormesis. Health Physics 52: 517.
32. J. J. Cohen, 1987Conference on radiation hormesis- Overview. Health Physics 52:519.
33. Z. L. Shu, W. H. Liu, J. B. Sun, 1987Radiation hormesis- Its expression in the immune system. Health Physics 52579583
34. M. W. Miller, 1987Radiation hormesis in plants.Health Physics52607616
35. P. Weis, and J. S. Weis, 1986Cadmium acclimation and hormesis in Fundulus heteroclitus during fin regeneration.Environmental Research39356363
36. A. Furst, 1987Hormetic Effects in pharmacology- Pharmacological inversions as prototypes for hormesis. Health Physics 52527530
37. E. J. Calabrese, M. E. Mccarthy, and E. Kenyon, 1987The occurrence of chemically induced hormesis.Health Physics52531541
38. L. A. Sagan, 1989On radiation, paradigms, and hormesis. Science 245: 574.
39. E. J. Calabrese, and K. J. Howe, 1976Stimulation of growth of peppermint (Mentha piperita) by phosfon, a growth retardantPhysiologia Plantarum37163165
40. E. J. Calabrese, and L. A. Baldwin, 1993Possible examples of chemical hormesis in a previously published study.Journal of Applied Toxicology 13169172

41. E. J. Calabrese, and L. A. Baldwin, 1996Expanding the RfD concept to incorporate and optimize beneficial effects while preventing toxic responses from nonessential toxicants.Ecotoxicology and Environmental Safety3494101

42. E. J. Calabrese, and L. A. Baldwin, 1997a. A. Quantitatively-based, methodology for the evaluation of chemical hormesis. Human and Ecological Risk Assessment4545554

43. E. J. Masoro, 2005Overview of caloric restriction and ageingMechanisms of Ageing and Development126913922

44. L. L. Ji, M. C. Gomez-cabrera, J. Vina, 2006Exercise and Hormesis Activation of Cellular Antioxidant Signaling Pathway. In: Understanding and modulating aging. Annals of the New York Academy of Sciences1067425435

45. D. P. Hayes, 2007Nutritional hormesis.European Journal of Clinical Nutrition61147159

46. M. Ristow, K. Zarse, A. Oberbach, N. Kloting, M. Birringer, M. Kiehntopf, M. Stumvoll, C. R. Kahn, and M. Bluher, 2009Antioxidants prevent health-promoting effects of physical exercise in humansProceedings of the National Academy of Sciences of the United States of America 10686658670

47. 40921022113Tlili, A., Montuelle, B., Berard, A., and Bouchez, A. 2011. Impact of chronic and acute pesticide exposures on periphyton communities. Science of the Total Environment 409: 2102-2113

48. E. J. Calabrese, and R. B. Blain, 2009Hormesis and plant biology. Environmental Pollution 14248

49. E. D. Velini, E. Alves, M. C. Godoy, D. K. Meschede, R. T. Souza, S. O. Duke, 2008Glyphosate applied at low doses can stimulate plant growth.Pest Management Science64489496

50. L. Migliore, S. Cozzolino, M. Fiori, 2003Phytotoxicity to and uptake of enrofloxacin in crop plantsChemosphere 5212331244

51. B. D. Roebuck, Yager Jr. JD, Longnecker DS. 1981Dietary modulation of azaserine-induced pancreatic carcinogenesis in the rat.Cancer Reserach 41888893

52. J. B. Johnson, W. Summer, R. G. Cutler, B. Martin, D. H. Hyun, V. D. Dixit, M. Pearson, M. Nassar, R. Tellejohan, S. Maudsley, O. Carlson, S. John, D. R. Laub, M. P. Mattson, 2007Alternate day

calorie restriction improves clinical findings and reduces markers of oxidative stress and inflammation in overweight adults with moderate asthmaFree Radical Biology Medicine 42665674

53. N. Kinoshita, K. Sueki, K. Sasa, J. Kitagawa, S. Ikarashi, T. Nishimura, Y. S. Wong, Y. Satou, K. Handa, T. Takahashi, M. Sato, and T. Yamagata, 2011Assessment of individual radionuclide distributions from the Fukushima nuclear accident covering central-east Japan.Proceedings of the National Academy of Sciences of the United States of America1081952619529
54. R. N. C. Guedes, L. C. Magalhaes, and L. V. Cosme, 2009Stimulatory sublethal response of a generalist predator to permethrin: hormesis, hormoligosis, or homeostatic regulation?Journal of Economic Entomology102170176
55. M. Hotchkiss, 1923Studies on salt action. Journal of Bacteriology. 8141162In Stebbing, A. R. (1982). Hormesis- The stimulationof growth by low levels of inhibitors. The Science of the Total Environment 3:213-234.
56. W. S. Miller, C. A. Green, and H. Kitchen, 1945Biphasic action of Penicillin and other sulphonamide similarityNature1210211
57. J. F. Linares, I. Gustafsson, F. Baquero, J. L. Martinez, 2006Antibiotics as intermicrobial signaling agents instead of weaponsProceedings of the National Academy of Sciences of the United States of America 1031948419489
58. J. X. Wang, P. Xie, N. Guo, 2007Effects of nonylphenol on the growth and microcystin production of Microcystis strainsEnvironmental Research1037078
59. Y. Gong, L. R. Song, X. Q. Wu, B. D. Xiao, T. Fang, J. T. Liu, 2009Effects of arsenate on microcystin content and leakage of Microcystis Strain PCC7806 under various phosphate regimes. Environmental Toxicology248794
60. Y. Hong, H. Hong-ying, F. M. Li, 2008Growth and physiological responses of freshwater green alga Selenastrum capricornutum to allelochemical ethyl 2-methyl acetoacetate (EMA) under different initial algal densitiesPesticide Biochemistry and Physiology90203212
61. E. J. Calabrese, J. W. Staudenmayer, E. J. Stanek, 2006Hormesis outperforms threshold model in National Cancer Institute

antitumor drug screening database.Toxicological Sciences 94368378

62. A. Mesquita, M. Weinberger, A. Silva, B. Sampaio-marques, B. Almeida, C. Leao, V. Costa, F. Rodrigues, W. C. Burhans, P. Ludovico, 2010Caloric restriction or catalase inactivation extends yeast chronological lifespan by inducing H2O2 and superoxide dismutase activityProceedings of the National Academy of Sciences of the United States of America 1071512315128
63. C. Stevens, V. A. Khan, J. Y. Lu, C. L. Wilson, P. L. Pusey, M. K. Kabwe, E. C. K. Igwegbe, E. Chalutz, and S. Droby, 1998The germicidal and hermetic effects of UV-C light on reducing brown rot disease and yeast microflora of peaches. Crop Protection 177584
64. D. G. Hessayon, 1951Double-action of trichothecin and its production in soil.Nature4284998999
65. D. G. Hessayon, 1953Fungitoxins in the soil: II Trichothecin, its production and inactivation in unsterilized soils. Soil Science 5395404
66. G. W. Moorman, 2011Sublethal doses of mefenoxam enhance Pythium damping-off of geraniumPlant Disease9512331238
67. E. Baraldi, M. Mari, E. Chierici, M. Pondrelli, P. Bertolini, and G. C. Pratella, 2003Studies on thiabendazole resistance of Penicillium expansum of pears: pathogenic fitness and genetic characterizationPlant Pathology52362370
68. K. Audenaert, E. Callewaert, M. Höfte, S. De Saeger, and G. Haesaert, 2010Hydrogen peroxide induced by the fungicide prothioconazole triggers deoxynivalenol (DON) production by Fusarium graminearumBMC Microbiology 10:112.
69. J. Landry, C. Martinez, and L. Rochefort, 2011The use of fungicide Nova to mitigate infection of Sphagnum by parasitic fungi in the greenhouseBotany89655661
70. M. E. Fenn, and M. D. Coffey, 1984Studies on the in vitro and in vivo antifungal activity of fosetyl-Al and phosphorous acidPhytopathology74606611
71. S. Kato, R. Coe, L. New, and M. Dick, 1990Sensitivities of various Oomycetes to hymexazol and metalaxylJournal of General Microbiology13621272134

72. S. Zhang, S. G. Panaccione, M. E. Gallegly, 1997Metalaxyl stimulation of growth of isolates of Phytophthora infestansMycologia89289292
73. G. W. Moorman, and S. H. Kim, 2004Species of Pythium from greenhouses in Pennsylvania exhibit resistance to propamocarb and mefenoxamPlant Disease 88630632
74. F. Flores, and C. D. Garzon, 1 EOF13 EOF2012Detection and assessment of chemical hormesis on the radial growth in vitro of oomycetes and fungal plant pathogens.Dose-ResponseIn press.
75. E. J. Calabrese, 2008Hormesis and Medicine. British Journal of Clinical Pharmacology 66594617
76. J. M. Davies, C. V. Lowry, K. J. Davies, 1995Transcient adaptation to oxidative stressing yeast. Archives of Biochemestry and Biophysics 31716
77. J. A. Matchett, and C. K. Erickson, 1977Alteration of ethanolinduced changes in locomotor activity by adrenergic blockers in mice. Psychopharmacology 52201206
78. H. G. Baumgarten, A. Lachenmayer, A. Bjorklund, A. Nobin, and E. Rosengren, 1973Long-term recovery of serotonin concentrations in the rat CNS following 5,6- dihydroxytryptamineLife Sciences12357364
79. A. R. Stebbing, 2009Interpreting 'Dose-Response' curves using homeodynamic data: with an improved explanation for hormesis. Dose-Response 7221233
80. R. B. Conolly, and W. K. Lutz, 2004Nonmonotonic dose-response relationships: Mechanistic basis, kinetic modeling, and implications for risk assessmentToxicological Sciences1151157
81. G. Gómez-icazbalceta, L. Huerta, L. E. Soto-ramirez, and C. Larralde, 2007Extracellular HIV 1 Nef protein modulates lytic activity and proliferation of human CD8+ T lymphocytes.Cellular Immunology85 EOF90 EOF
82. M. Kushida, T. Sukata, S. Uwagawa, K. Ozaki, A. Kinoshita, H. Wanibuchi, K. Morimura, Y. Okuno, and S. Fukushima, 2005Low dose DDT inhibition of hepatocarcinogenesis initiated by diethylnitrosamine in male rats: Possible mechanismsToxicology and Applied Pharmacology3285294

83. O. N. Bae, K. M. Lim, J. Y. Han, B. I. Jung, J. Y. Lee, J. Y. Noh, S. M. Chung, M. Y. Lee, and J. H. Chung, 2008U-s. h. a. p. e. d. Dose-response, in vasomotor tone: A mixed result of heterogenic response of multiple cells to xenobiotics. Toxicological Sciences1181190
84. W. J. Allender, 1997Effect of trifluoperazine and verapamil on herbicide stimulated growth of cottonJournal of Plant Nutrition16980
85. W. J. Allender, G. C. Cresswell, J. Kaldor, and I. R. Kennedy, 1997Effect of lithium and lanthanum on herbicide induced hormesis in hydroponically-grown cotton and cornJournal of Plant Nutrition18195
86. K. Crump, 2001Evaluating the evidence for hormesis: A statistical perspective.Human and Ecological Risk Assessment 4:781.
87. G. A. Mack, and D. A. Wolfe, 1981K-Sample Rank Tests for Umbrella AlternativesJournal of the American Statistical Association76175181
88. H. Buning, and W. Kossler, 1997Power of some tests for umbrella alternatives in the multi-sample location problemBiometrical Journal39481494
89. R. V. Hogg, D. M. Fisher, and R. H. Randles, 1975A. Two-sample, Adaptive Distribution-Free Test. Journal of the American Statistical Association70656661
90. J. L. Gastwirth, 1965Percentile Modifications of Two Sample Rank Tests.Journal of the American Statistical Association 6031211271141
91. R. E. Barlow, 1972Statistical inference under order restrictions: the theory and application of isotonic regressionJ. Wiley, 388p.
92. A. J. Bailer, and J. T. Oris, 1997Estimating inhibition concentrations for different response scales using generalized linear modelsEnvironmental Toxicology and Chemistry1615541559
93. R. J. Hickey, E. J. Bowers, and R. C. Clelland, 1983Radiation hormesis, public health, and public policy: a commentary. Health Phys 44320719
94. H. Boxenbaum, P. J. Neafsey, and D. J. Fournier, 1988Hormesis, Gompertz functions, and risk assessment. Drug Metab Rev 19195229

95. P. J. Neafsey, H. Boxenbaum, D. A. Ciraulo, and D. J. Fournier, 1988A Gompertz age-specific mortality rate model of aging, hormesis, and toxicity: fixed-dose studies.Drug Metab Rev 19369401

96. P. Brain, and R. Cousens, 1989An equation to describe dose-responses where there is stimulation of growth at low dosesWeed Research299396

97. C. Q. Deng, R. Graham, and R. Shukla, 2001Detecting and estimating hormesis using a model-based approachHuman and Ecological Risk Assessment74849866

98. P. H. Van Ewijk, and J. A. Hoekstra, 1993Calculation of the EC50 and its confidence interval when subtoxic stimulus is present. Ecotoxicology and Environmental Safety252532

99. N. Cedergreen, C. Ritz, and J. C. Streibig, 2005Improved empirical models describing hormesis.Environmental Toxicology and Chemistry 241231663172

100. J. M. Albourie, J. Tourvieille, and D. T. De Labrouhe, 1998Resistance to metalaxyl in isolates of the sunflower pathogen Plasmopara halstediiEuropean Journal of Plant Pathology104235242

101. K. H. Lamour, and M. K. Hausbeck, 2003Susceptibility of mefenoxam-treated cucurbits to isolates of Phytophthora capsici sensitive and insensitive to mefenoxamPlant Disease, 87920922

102. G. W. Moorman, S. Kang, D. M. Geiser, and S. H. Kim, 2002Identification and characterization of Pythium species associated with greenhouse floral crops in PennsylvaniaPlant Disease, 8612271231

103. R. J. Taylor, B. Salas, G. A. Secor, V. Rivera, and N. C. Gudmestad, 2002Sensitivity of north American isolates of Phytophthora erythroseptica and Pythium ultimum to mefenoxam (metalaxyl) Plant Disease, 86797802

104. N. J. Grunwald, A. K. Sturbaum, G. R. Montes, E. G. Serrano, H. Lozoya-saldana, and W. E. Fry, 2006Selection for fungicide resistance within a growing season in field populations of Phytophthora infestans at the center of origin. Phytopathology9613971403

105. U. Gisi, and H. Sierotzki, 2008Fungicide modes of action and resistance in downy mildewsEuropean Journal of Plant Pathology122157167
106. FRAC 2011FRAC list of plant pathogenic organisms resistant to disease control agents. Fungicide Resistance Action Committee, 71p.
107. G. W. Moorman, and R. J. Lease, 1992Benzimidazole- and dicarboximide-resistant Botrytis cinerea from Pennsylvania greenhouses. Plant Dis. 76477480

Chapter 5

Fungicide: Modes of Action and Possible Impact on Nontarget Microorganisms

Chao Yang[1,2], Chantal Hamel[1], Vladimir Vujanovic[2], and Yantai Gan[1]

[1]Semiarid Prairie Agricultural Research Centre, AAFC, Swift Current, SK, Canada S9H 3X2

[2]Department of Food and Bioproducts Sciences, University of Saskatchewan, Saskatoon, SK, Canada S7N 5N8

ABSTRACT

Fungicides have been used widely in order to control fungal diseases and increase crop production. However, the effects of fungicides on microorganisms other than fungi remain unclear. The modes of action of fungicides were never well classified and presented, making difficult to estimate their possible nontarget effects. In this paper, the action modes and effects of fungicides targeting cell membrane components, protein synthesis, signal transduction, respiration, cell mitosis, and nucleic acid synthesis were classified, and their effects on nontarget microorganisms were reviewed. Modes of action and potential non-target effects on soil microorganisms should be considered in the selection of fungicide in order to protect the biological functions of soil and optimize the benefits derived from fungicide use in agricultural systems.

INTRODUCTION

Soil is arguably the most important resource for food production. It is a very complex system whose functions not only depend on its physical properties, but also on its biological components. In particular, soil microorganisms are essential players in the cycling of several elements essential to life, including C, N, and P [1].

Understanding the effect of fungicides on the beneficial activities of microorganisms is important to assess the hazards associated with fungicide used in agriculture. Crop productivity and economic returns will be maximized with the use of products controlling well fungal pathogens, but preserving beneficial organisms. Different organisms may possess identical or similar mechanisms and constituents, and fungicides targeting nonspecific binding sites can directly affect nontarget organisms. For example, the toxicity of carboxylic acid fungicides is derived from the ability of these chemicals to bind on DNA topoisomerase II, as common enzyme that unwind, and wind, DNA to allow protein synthesis and DNA replication. This enzyme is found in fungi but also in prokaryotic cells [2]. Some glucopyranosyl antibiotic fungicides are toxic to bacteria, in which they may inhibit the synthesis of amino acids [3]. These fungicides are also toxic to certain nonfungal higher eukaryotic organisms [4].

Indirect nontarget effects are also possible. Microorganisms are either functionally or nutritionally connected with each other's, and changes in a component of a microbial community may influence the structure of the whole community. This is particularly true for plant-associated microorganisms, which influence on and are influenced by the plant metabolic status [5–7].

In order to establish a proper regulation for the use of the many fungicidal substances promoted by industry in sustainable agriculture, fungicide action modes and possible side effects on nonfungal microorganisms must urgently be clarified. Fungicide action modes have never been well classified, and the side effects of these important chemicals are not fully understood. Therefore, fungicide use may have negative impacts that are difficult to predict [8]. In this paper, current knowledge on the action modes of fungicides impacting membranes, nucleic acids and protein synthesis, signal transduction, respiration, mitosis and cell division, and Multisite activity, as well as on their side effects on nontarget organisms will be summarized and organized. The framework emerging from this analysis sheds a much needed light on the possible side effects of the numerous fungicidal products in use and facilitates the assessment of the risks associated with their use. The information summarized here will support the development of efficient agroecosystems where the contribution of naturally occurring bioresources is preserved.

MODES OF ACTION AND SIDE EFFECTS OF FUNGICIDE GROUPS

Effects on the Synthesis of Lipids, Sterol, and Other Membrane Components

The cell membrane is a selectively-permeable wall that separates the cell content from the outside environment. Membranes perform many biological functions in all living cells. They preclude the passage of large molecules, provide the shape of the cell, maintain cell water potentials, and are involved in signal transduction [9]. Negative impacts of fungicide on the membrane of microorganisms were found to alter the structure and function of soil microbial communities.

The structure of lipids, the basic components of cell membranes, was modified by fungicides of the Aromatic Hydrocarbons (AH) group, impacting the functionality of microbial membrane systems. For instance, Dicloran (2,6-dichloro-4-nitroaniline)—an AH fungicide registered in North America, Europe, and South Africa since 1975 for the control of Basidiomycetes, Deuteromycetes, and Rhizopus species [10]—is phototoxic. The cell membranes of treated fungi become sensitive to solar radiation, which then destroys the structure of linoleic acid, a common membrane lipid. Another active AH fungicide ingredient, etridiazole (5-ethoxy-3(trichloromethyl)-1,2,4-thiadiazole), causes the hydrolysis of cell membrane phospholipids into free fatty acids and lysophosphatides [11], leading to the lysis of membranes, in fungi. Previous research proved that these fungicides have side effects on other soil microorganisms. Dicloran can cause mutation in Salmonella typhimurium by disturbing hydrophobic interactions within the membrane [12]. Etridiazole also reduced the nitrification rate of ammonium-oxidizing bacteria in soil [13], with possible effect on this component of the soil microbial community and ramifications on its structure and function.

Sterols are another important component of cell membrane in fungi. Demethylation-inhibiting (DMI) fungicides inhibit sterol biosynthesis in fungal cells. Triadimefon ((RS)-1-(4-chlorophenoxy)-3,3-dimethyl-1-(1H-1,2,4-triazol-1-yl)butan-2-one) demethylated at C-14, introduced a double bond at C-22, and reduced a double bond at C-24 in the carbon skeleton of sterols in a fungal membrane, causing disfunction and cell lysis [14]. Although bacteria do not have sterols, sterol-targeting fungicides have indirect side effects on these microorganisms. Research found that triadimefon had long-term inhibiting effects on soil bacterial community [7]. Triticonazole ((RS)-(E)-5-(4-chlorobenzylidene)-2,2-dimethyl-1-(1H-1,2,4-triazol-1 ylmethyl)cyclopentanol), a triazoles fungicide, can stimulate bacteria proliferation in soil [15], while two other sterol-targeting fungicides, fenpropimorph (cis-4-[(RS)-3-(4-tert-butylphenyl)-2-methylpropyl]-2,6-dimethylmorpholine) and propiconazole ((2RS,4RS;2RS,4SR)-1-[2-(2,4-dichlorophenyl)-4-propyl-1,3-dioxolan-2-ylmethyl]-1H-1,2,4-triazole), inhibited overall bacterial activity [16]. Such differential effect may be explained by changes in competition among different soil microorganisms. Recent research of dimethomorph ((EZ)-4-[3-(4-chlorophenyl)-3-(3,4-dimethoxyphenyl)acryloyl]morpholine) revealed that this fungicide can influence the activity of bacteria involved in N cycling, with impact on nitrification and ammonification [17], through its different impact on different bacterial ecotypes and changes in bacterial community structure.

Some fungicides target fungal intracellular membrane systems and their biological functions. A widely used fungicidal compound, acriflavine (3,6-diamino-10-methylacridin-10-ium chloride), increases mitochondrial permeability and releases cytochrome c in fungal cells, repressing plasma membrane receptor activation, disordering proton stream and collapsing the electrochemical proton gradient across mitochondrial membranes [18]. As a consequence, ATP synthesis is decreased leading to cell death. It was also shown that acriflavine could thickened both the peripheral and cross cell wall of the gram-negative bacteria Staphylococcus aureus [18], suggesting the possibility of nontarget effects of acriflavine on bacterial growth (Table 1).

Table 1: Action mode and possible nontarget effects of fungicides

Action mode		Fungicide chemical group	Common name	Nontarget effects
Lipid, sterol, and other membrane components	Lipid	Aromatic hydrocarbons	Dicloran	Mutagen to Salmonella typhimurium [12]
			Etridiazole	Retards nitrification by affecting ammonium oxidizers [13]
	Sterol	Triazoles	Triadimefon	Long-term inhibiting effects on soil bacterial community [7]
			Triticonazole	Increases total number of bacteria in soil [15]
		Cinnamic acid amide	Dimethomorph	Impacts nitrifying and ammonifying bacterial activities in sandy soils [17]
		Triazole	Hexaconazole	Impacts bacterial activities related to N cycling [19]
		Morpholine	Fenpropimorph	Inhibit general bacterial activity in wetland [16]
		Triazole	Propiconazole	
			Tebuconazole	May retard plant-growth-promoting effects ofAzospirillum brasilense on its hostplant [20]
	Intracellular membrane components	Hydrochloride	Acriflavine	Thickens peripheral and cross cell wall ofStaphylococcus aureus [18]

Amino acid and protein synthesis		Glucopyranosyl antibiotic	Streptomycin	Inhibits amino acid synthesis in bacteria [3] and is neurotoxic to amphibian [4]
		Tetracycline antibiotic	Oxytetracycline	Also used as bactericide [21]
Signal transduction		Phenylpyrroles	Fludioxonil	Toxic to algae [22] and potential risk to prokaryotes [23]
		Dicarboximides	Iprodione	Affects signal transduction in bacteria [24]
			Vinclozolin	Inhibits total bacterial growth [25]
Respiration	NADH oxido-reductase (Complex I) inhibitors	Pyrimidinamines	Diflumetorim	Unknown
	Succinate-dehydrogenase (Complex II) inhibitors	Pyridine carboxamides	Boscalid	May affect growth of prokaryotes [26]
		Benzamides	Flutolanil	
		Gxathiin carboxamides	Carboxin	Inhibits denitrifying bacterial activity in wetland sediment [16]
	Oxidative phosphorylation uncouplers	2,6-dinitroanilines	Fluazinam	Have a potential risk to environmental microorganisms [27]
		Dinitrophenyl crotonate	Dinocap	Inhibits ammonifying bacterial activity and stimulate general bacterial respiration in soil [28]

Mitosis and cell division	Inhibitor of spindle microtubules assembly	Methyl benzimidazole carbamate	Benomyl	May affect nitrifying bacteria [29] and arbuscular mycorrhizal fungi [30]
			Carbendazim	Reduces the diversity of soil bacteria [31]
		Phenylurea	Pencycuron	May affect metabolically activated soil bacteria in short term [32]
Nucleic acids synthesis	RNA polymerase I inhibitors	Acylalanines	Metalaxyl	Affects activities of ammonifying and nitrifying bacteria in soil [33]
		Oxazolidinones	Oxadixyl	Unknown
	Adenosin-deaminase inhibitors	Hydroxypyrimidines	Ethirimol	Unknown
		Phthalonitrile	Chlorothalonil	Impacts bacterial activities related to N cycling [29]
Multisite activity		Dithiocarbamate	Mancozeb	Impacts bacterial activities related to nitrogen cycling [17] and carbon cycling [28] in soils
		Phthalimide	Captan	Inhibits denitrifying bacterial activity [16]
		Dithiocarbamate	Thiram	
		Anthraquinone	Dithianon	Reduces bacterial diversity in soil [34]
		Copper	Copper sulfate	Reduces the number of bacteria and streptomycetes in sandy soil [35]

Effects on Amino Acids and Protein Synthesis

Proteins are the most important building blocks in living organisms. They have various important biological functions such as making up the cytoskeleton, delivering signals among cells, and catalyzing biochemical reactions [36]. Proteins are made of amino acids. Several fungicides interfere with the biosynthesis of amino acids and proteins, affecting the biological functions of impacted organisms.

Streptomycin (5-(2,4-diguanidino-3,5,6-trihydroxy-cyclohexoxy)-4-[4,5-dihydroxy-6-(hydroxymethyl)-3-methylamino-tetrahydropyran-2-yl]oxy-3-hydroxy-2-methyl-tetrahydrofuran-3-carbaldehyde), an antibiotic produced by Streptomyces griseus that has long been used as a fungicide [37], also has bactericidal activity. Streptomycin interferes with amino acid synthesis. In Escherichia coli, application of streptomycin caused misincorporation of an isoleucine molecule in the phenylalanine polypeptide chain associated with 70S ribosomes [38]. Another research with a thermus thermophilus mutant strain suggested that misreading of the genes coding for amino acid synthesis explains the negative effect of streptomycin on bacteria [3]. Furthermore, Perez et al. [4] found that streptomycin could also be a nonselective excitatory amino acid (EAA) receptor antagonist. This antibiotic selectively blocked amino acid receptors in the anterior vestibular nerve fibers ofAmbystoma tigrinum, a salamander, suggesting that it could also be toxic to eukaryotes, in addition to fungi and bacteria.

Oxytetracycline ((2Z,4S,4aR,5S,5aR,6S,12aS)-2-[amino(hydroxy) methylidene]-4-(dimethylamino)-5,6, 10,11, 12a-pentahydroxy-6-methyl-4,4a,5,5a-tetrahydrotetracene-1,3,12-trione) is widely used in agriculture because of its broad-spectrum antibiotic activity. It is also registered as fungicide in New Zealand and VietNam, according to the information provided by Pesticide Action Network of North America (http://www.pesticideinfo.org/Detail_ChemReg.jsp?Rec_Id=PC38140). Previous research reported inhibitory effects of oxytetracycline on protein synthesis in bacteria through interference with the ternary amino-acyl-tRNA complex binding to the acceptor site of ribosomes [39], leading to retarded bacterial growth, disordered microbial community structure, and limited microbial ectoenzyme activity in the soil system [21, 40]. Therefore, caution must be taken with the application of oxytetracycline to control fungal diseases, as it is antibiotic and impacts bacteria.

Effects on Signal Transduction

The fungicide affecting microbial membranes or proteins, as we discussed above, may affect signal transduction, which takes place at the level of membranes and involves the function of certain proteins.

Phenylpyrrole fungicidal ingredient fludioxonil (4-(2,2-difluoro-1,3-benzodioxol-4-yl)-1H-pyrrole-3-carbonitrile), is a nonsymtemic fungicide, also known to interfere with the signal transduction pathways of target fungi [41]. The work of Rosslenbroich and Stuebler [42] revealed inhibition of spore germination, germ tube elongation, and mycelium growth in Botrytis cinerea, by fludioxonil-related interference in the osmoregulatory signal transmission pathway of this fungus. This finding was supported by Ochiai et al. [43] who found that fludioxonil can disturb the CANIKI/COSI signal transduction pathway, leading to the dysfunction of glycerol synthesis and inhibition of hyphae formation in Candida albicans. Recently, Hagiwara et al. [44] reported the inhibiting effect of fludioxonil on a large number of genes involved in a two-component signal transduction system, in filamentous fungi. Impact on this system suggests that fludioxonil may have a nontarget effect on bacteria, as this dualistic signal transduction mechanisms is also reported in prokaryotes [45].

Effects on signal transduction are also found in dicarboximide fungicides. Iprodione (3-(3,5-dichlorophenyl)-N-isopropyl-2,4-dioxoimidazolidine-1-carboxamide), a contact dicarboximide fungicide widely used in a variety of crops, inhibits glycerol synthesis and hyphal development by cutting off signal transduction [43], as does fludioxonil. Iprodione can modify the structure of the soil bacterial community, as reported in a recent research [24]. Interference with signal transduction by dicarboximide fungicide vinclozolin ((RS)-3-(3,5-dichlorophenyl)-5-methyl-5-vinyl-1,3-oxazolidine-2,4-dione) caused low growth rate, abnormality, and changes in the productions of hexoses and chitin in treated B. cinerea [46]. Vinclozolin also had inhibiting effects on soil bacterial growth and nitrogen metabolism, in soil systems [25]. The metabolite of this fungicidal compound, 3,5-dichloroaniline, is also toxic and persistent [23], further suggesting possible impacts of the fungicide vinclozolin on nontarget soil organisms.

Effects on Respiration

Several fungicides with different modes of action were reported to inhibit microbial respiration. Some are NADH oxidoreductase (Complex I) inhibitors, others are succinate-dehydrogenase (Complex II) inhibitors, cytochrome bc1 (Complex III) inhibitors, and oxidative phosphorylation uncouplers.

Only few fungicides were reported so far to inhibit respiration by affecting Complex I system in fungal mitochondria. Diflumetorim ((RS)-5-chloro-N-{1-[4-(difluoromethoxy)phenyl]propyl}-6-methylpyrimidin-4-ylamine), first registered in Japan in 1997 to control powdery mildew and rust in ornamental plants [47], inhibits NADH oxido-reductase activity leading to fungal death [48]. Very limited research has investigated the mode of action mode of Complex I inhibitors, which remains poorly understood.

Three widely used Complex II inhibitors, boscalid (2-chloro-N-(4′-chlorobiphenyl-2-yl) nicotinamide), carboxin (5,6-dihydro-2-methyl-1,4-oxathiine-3-carboxanilide), and flutolanil(α,α,α-trifluoro-3′-isopropoxy-o-toluanilide), cause dysfunction of succinate dehydrogenase (SDH) in the tricarboxylic cycle and mitochondrial electron transport chain, inhibiting the activity of Complex II and respiration in fungal cells [49–52]. Significant yield increases were reported with the use of these fungicides, indicating their effectiveness in the control of fungal diseases [53, 54]. Since Complex II is a common enzyme complex system existing in many eukaryotic and prokaryotic organisms [26], nontarget effects of Complex II inhibitors on soil bacteria were repeatedly reported [55, 56], suggesting that cautions must be used with these chemicals.

Whereas some fungicides affect fungal respiration at the level of the enzyme complex system, other fungicides may impact respiration through other targets. Fluazinam (3-chloro-N-(3-chloro-5-trifluoromethyl-2-pyridyl)-α,α,α-trifluoro-2,6-dinitro-p-toluidine) triggers very unusual uncoupling activity in target cells. The metabolic state of their mitochondria was found to be inhibited after exposure to fluazinam, which may be caused by the conjugation of the chemical with glutathione, in mitochondria [57]. Consequently, ATP production is inhibited and downstream cellular metabolisms is interrupted. In fact, the uncoupling activity of eight fluazinam derivatives was recognized

[58], which suggests that fluazinam has complicated ramifications on fungal metabolic pathways and may be toxic in the environment [27]. Another fungicide dinocap (RS)-2,6-dinitro-4-octylphenyl crotonates and (RS)-2,4-dinitro-6-octylphenyl) showed similar action mode to fluazinam, which inhibited ammonifying bacterial activity [28], suggesting side effects of this fungicide group on bacteria growth.

Effects on Mitosis and Cell Division

The methyl benzimidazole carbamate (MBC) fungicides are known to impact mitosis and cell division in target fungi [59, 60]. Previous research revealed the inhibitory effects of these fungicides on the polymerization of tubulin into microtubules. These MBC fungicides bind on β-tubulin in microtubules inhibiting their proliferation and suppressing their dynamic instability [61–63]. Microtubules are the cytoskeletal polymers in eukaryotic cells and, thus, play a vital role in many cellular functions. The application of MBC fungicides suppresses the assembly of spindle microtubules, disturbs the chromosomal alignment at the metaphase plate and microtubule-kinetochore interacions causing chromatid loss, chromosome loss or nondisjunction in target cells [64], which may also yield side effects on other microorganisms as described below.

Benomyl (methyl 1-(butylcarbamoyl) benzimidazol-2-ylcarbamate) and carbendazim (methyl benzimidazol-2-ylcarbamate), two very popular MBC fungicides widely used in crop production, inhibit mitosis in fungi. They can also influence the beneficial arbuscular mycorrhiza fungi (AMF) [30] and mammalian cells [65, 66]. Although no evidence of a direct effects of MBC fungicides on soil bacteria was reported yet, some research has associated these fungicides to the inhibition of nitrification in soil, a microbially mediated process [29]. The effect of MBC fungicides on bacteria and other soil organisms remains to be clarified.

Effects on Nucleic Acids Synthesis

Phenylamides (PA) fungicides affect nucleic acids synthesis by inhibiting the activity of the RNA polymerase I system. For example, metalaxyl (methyl N-(methoxyacetyl)-N-(2,6-xylyl)-DL-alaninate), a widely used PA fungicide, inhibits uridine incorporation into the RNA chain [67]. It interferes with nucleic acid synthesis through inhibition of RNA polymerase I activity thus blocking rRNA synthesis at the level of uridine transcription [68]. PA fungicide applications can increase the prevalence of fungicide resistance in pathogen population and yield more fungicide-resistant isolates, as shown by a recent study using AFLP (amplified fragment length polymorphism) and SSR (simple sequence repeats) markers [69]. Fungicides in the PA group must be used with caution, as the side effect of this fungicide on N cycling associated bacteria was reported [33].

Hydroxypyrimidines fungicides were also reported for their inhibiting effects on adenosine-deaminase. As an example, ethirimol (5-butyl-2-ethylamino-6-methylpyrimidin-4-ol) was reported for its effects on several metabolites such as inosine and adenine nucleotides in barley powdery mildew (Erysiphe graminis f.sp. hordei.) [70]. Ethirimol caused overexpression of adenine phosphoribosyltransferase, which may further break down the balance of the nucleotide pool. Besides, ADAase, which catalyzes the hydrolytic deamination of adenosine, was inhibited by ethirimol. Consequently, production of inosine was ceased, and synthesis of nucleic acid was impaired. The gene responsible for resistance to ethirimol, ethIS, was reported later in Erysiphe graminis f.sp.hordei [71]; therefore, caution must be taken with the application of hydroxypyrimidines fungicide as fungicide resistance in target populations could be developed by repeatedly fungicide application.

Fungicides with Multisite Activity

Multisite activity fungicides are widely used in agronomic activities due to the broad spectrum of disease control activity, but may have side effects on other microorganisms due to their multiple biochemical sites impacts. Chlorothalonil (tetrachloroisophthalonitrile), a widely used phthalonitrile fungicide, can block the transformation of alternative special structure of glutathione and reduce enzymes activities which used special conformation of glutathione as their reaction centers. Previous research found that chlorothalonil can influence bacterial growth in soil, which may have ecological consequences on N cycling [29]. Mancozeb (manganese ethylenebis(dithiocarbamate) (polymeric) complex with zinc salt), another Multisite activity fungicide impacting metabolism in target cells, can also affect bacteria involved in both C and N cycling in soil [17, 28]. Other Multisite activity fungicides such as captan (N-cyclohex(trichloromethylthio)-4-ene-1,2-dicarboximide) and thiram (bis(dimethylthiocarbamoyl) disulfide) inhibited the growth of denitrifying bacteria [16], perhaps due to their nonspecific effects on biochemical compounds which contain thiol in target cells. Besides, copper-based Multisite activity fungicide, such as copper sulfate (copper(II) tetraoxosulfate), inhibited bacteria and streptomycetes growth in soil [35] and may have nontarget effects on other soil microorganisms.

CONCLUSIONS

Fungicidal compounds may have side effects and impact nontarget soil microorganism. The effects of fungicides on soil microorganisms can be important, as the feedback of the soil microbial community can affect crops growth and production in cropping systems. The relationships existing between fungicides, the soil microorganisms, and other environmental factors are complex and difficult to predict. On the other hand, the multiplicity of fungicides' modes of action increases the difficulty of evaluating the risks associated with fungicide use. Since it is desirable to optimize the benefit of natural soil biological functions to crop production, understanding fungicides mode of action and impact on metabolism could help us using fungicide more wisely in agriculture.

ACKNOWLEDGMENTS

The authors gratefully acknowledge the financial support of Novozymes, Saskatchewan Pulse Growers, and Agriculture and Agri-Food Canada Matching Investment Initiative.

REFERENCES

1. C. Aponte, T. Marañón, and L. V. García, "Microbial C, N and P in soils of Mediterranean oak forests: influence of season, canopy cover and soil depth," Biogeochemistry, vol. 101, no. 1, pp. 77–92, 2010.
2. M. Sioud, A. Boudabous, and L. Cekaite, "Transcriptional responses of Bacillus subtillis and thuringiensis to antibiotics and anti tumour-drugs," International Journal of Molecular Medicine, vol. 23, no. 1, pp. 33–39, 2009.
3. J. F. Carr, S. T. Gregory, and A. E. Dahlberg, "Severity of the streptomycin resistance and streptomycin dependence phenotypes of ribosomal protein S12 of Thermus thermophilus depends on the identity of highly conserved amino acid residues," Journal of Bacteriology, vol. 187, no. 10, pp. 3548–3550, 2005.
4. M. E. Perez, E. Soto, and R. Vega, "Streptomycin blocks the postsynaptic effects of excitatory amino acids on the vestibular system primary afferents," Brain Research, vol. 563, no. 1-2, pp. 221–226, 1991.
5. P. M. White, T. L. Potter, and A. K. Culbreath, "Fungicide dissipation and impact on metolachlor aerobic soil degradation and soil microbial dynamics," Science of the Total Environment, vol. 408, no. 6, pp. 1393–1402, 2010.
6. Y. S. Wang, C. Y. Wen, T. C. Chiu, and J. H. Yen, "Effect of fungicide iprodione on soil bacterial community," Ecotoxicology and Environmental Safety, vol. 59, no. 1, pp. 127–132, 2004.
7. J. H. Yen, J. S. Chang, P. J. Huang, and Y. S. Wang, "Effects of fungicides triadimefon and propiconazole on soil bacterial communities," Journal of Environmental Science and Health Part B, vol. 44, no. 7, pp. 681–689, 2009.
8. C. C. Lo, "Effect of pesticides on soil microbial community," Journal of Environmental Science and Health Part B, vol. 45, pp. 348–359, 2010.
9. B. Alberts, A. Johnson, and J. Lewis, Molecular Biology of the Cell, Garland Science, New York, NY, USA, 4th edition, 2002.
10. F. Boscá, M. A. Miranda, G. Serrano, and F. Vargas, "Photochemistry and photobiological properties of dicloran, a postharvest

fungicide with photosensitizing side effects," Photochemistry and Photobiology, vol. 67, no. 5, pp. 532–537, 1998.

11. B. Radzuhn and H. Lyr, "On the mode of action of the fungicide etridiazole," Pesticide Biochemistry and Physiology, vol. 22, no. 1, pp. 14–23, 1984.
12. D. P. De Ouveira, M. Sakagami, S. Warren, F. Kummrow, and G. A. De Umbuzeiro, "Evaluation of dicloran's contribution to the mutagenic activity of cristais river, Brazil, water samples," Environmental Toxicology and Chemistry, vol. 28, no. 9, pp. 1881–1884, 2009.
13. G. A. Rodgers, "Potency of nitrification inhibitors following their repeated application to soil," Biology and Fertility of Soils, vol. 2, no. 2, pp. 105–108, 1986.
14. R. J. Pring, "Effects of triadimefon on the ultrastructure of rust fungi infecting leaves of wheat and broad bean (Vicia faba)," Pesticide Biochemistry and Physiology, vol. 21, no. 1, pp. 127–137, 1984.
15. A. Niewiadomska, Z. Sawińska, and A. W. Maruwka, "Impact of selected seed dressings on soil microbiological activity in spring barley cultivation," Fresenius Environmental Bulletin, vol. 20, pp. 1252–1261, 2011.
16. S. Milenkovski, E. Bååth, P. E. Lindgren, and O. Berglund, "Toxicity of fungicides to natural bacterial communities in wetland water and sediment measured using leucine incorporation and potential denitrification," Ecotoxicology, vol. 19, no. 2, pp. 285–294, 2010.
17. M. Cycoń, Z. Piotrowska-Seget, and J. Kozdrój, "Responses of indigenous microorganisms to a fungicidal mixture of mancozeb and dimethomorph added to sandy soils," International Biodeterioration and Biodegradation, vol. 64, no. 4, pp. 316–323, 2010.
18. M. Kawai and J.-I. Yamagishi, "Mechanisms of action of acriflavine: electron microscopic study of cell wall changes induced in Staphylococcus aureus by acriflavine," Microbiology and Immunology, vol. 53, no. 9, pp. 481–486, 2009.
19. R. J. Madhuri andV. Rangaswamy, "Influence of selected fungicides on microbial population in groundnut (Arachis hypogeae L.) soils," Pollution Research, vol. 22, no. 2, pp. 205–212, 2003.

20. M. A. Pereyra, F. M. Ballesteros, C. M. Creus, R. J. Sueldo, and C. A. Barassi, "Seedlings growth promotion by Azospirillum brasilense under normal and drought conditions remains unaltered in Tebuconazole-treated wheat seeds," European Journal of Soil Biology, vol. 45, no. 1, pp. 20–27, 2009.
21. Q. Yang, J. Zhang, K. Zhu, and H. Zhang, "Influence of oxytetracycline on the structure and activity of microbial community in wheat rhizosphere soil," Journal of Environmental Sciences, vol. 21, no. 7, pp. 954–959, 2009.
22. S. Verdisson, M. Couderchet, and G. Vernet, "Effects of procymidone, fludioxonil and pyrimethanil on two non-target aquatic plants," Chemosphere, vol. 44, no. 3, pp. 467–474, 2001.
23. J. B. Lee, H. Y. Sohn, K. S. Shin et al., "Microbial biodegradation and toxicity of vinclozolin and its toxic metabolite 3,5-dichloroaniline," Journal of Microbiology and Biotechnology, vol. 18, no. 2, pp. 343–349, 2008.
24. G. G. Miñambres, M. Y. Conles, E. I. Lucini, R. A. Verdenelli, J. M. Meriles, and J. A. Zygadlo, "Application of thymol and iprodione to control garlic white rot (Sclerotium cepivorum) and its effect on soil microbial communities," World Journal of Microbiology and Biotechnology, vol. 26, no. 1, pp. 161–170, 2010.
25. A. Banerjee and A. K. Banerjee, "Effect of the fungicides tridemorph and vinclozolin on soil microorganisms and nitrogen metabolism," Folia Microbiologica, vol. 36, no. 6, pp. 567–571, 1991.
26. K. S. Oyedotun and B. D. Lemire, "The quaternary structure of the Saccharomyces cerevisiae succinate dehydrogenase: homology modeling, cofactor docking, and molecular dynamics simulation studies,"Journal of Biological Chemistry, vol. 279, no. 10, pp. 9424–9431, 2004.
27. R. P. A. van Wijngaarden, G. H. P. Arts, J. D. M. Belgers et al., "The species sensitivity distribution approach compared to a microcosm study: a case study with the fungicide fluazinam," Ecotoxicology and Environmental Safety, vol. 73, no. 2, pp. 109–122, 2010.

28. J. Černohlávková, J. Jarkovský, and J. Hofman, "Effects of fungicides mancozeb and dinocap on carbon and nitrogen mineralization in soils," Ecotoxicology and Environmental Safety, vol. 72, no. 1, pp. 80–85, 2009.
29. S. K. Chen, C. A. Edwards, and S. Subler, "Effects of the fungicides benomyl, captan and chlorothalonil on soil microbial activity and nitrogen dynamics in laboratory incubations," Soil Biology and Biochemistry, vol. 33, no. 14, pp. 1971–1980, 2001.
30. G. R. Carr and M. A. Hinkley, "Germination and hyphal growth of Glomus caledonicum on water agar containing benomyl," Soil Biology and Biochemistry, vol. 17, no. 3, pp. 313–316, 1985.
31. X. Wang, M. Song, C. Gao et al., "Carbendazim induces a temporary change in soil bacterial community structure," Journal of Environmental Sciences, vol. 21, no. 12, pp. 1679–1683, 2009.
32. R. Pal, K. Chakrabarti, A. Chakraborty, and A. Chowdhury, "Pencycuron application to soils: degradation and effect on microbiological parameters," Chemosphere, vol. 60, no. 11, pp. 1513–1522, 2005.
33. A. Monkiedje and M. Spiteller, "Degradation of metalaxyl and mefenoxam and effects on the microbiological properties of tropical and temperate soils," International Journal of Environmental Research and Public Health, vol. 2, no. 2, pp. 272–285, 2005.
34. J. Liebich, A. Schäffer, and P. Burauel, "Structural and functional approach to studying pesticide side-effects on specific soil functions," Environmental Toxicology and Chemistry, vol. 22, no. 4, pp. 784–790, 2003.
35. O. Kostov and O. Van Cleemput, "Microbial activity of Cu contaminated soils and effect of lime and compost on soil resiliency," Compost Science and Utilization, vol. 9, no. 4, pp. 336–351, 2001.
36. H. Lodish, A. Berk, C. A. Kaiser, et al., Molecular Cell Biology, W. H. Freeman, London, UK, 4th edition, 2000.
37. B, M. D. Singh, "Bactericidal activity of streptomycin and isoniazid against tubercle bacilli," The British Medical Journal, vol. 1, pp. 130–132, 1954.

38. D. Old and L. Gorini, "Amino acid changes provoked by streptomycin in a polypeptide synthesized in vitro," Science, vol. 150, no. 3701, pp. 1290–1292, 1965.

39. D. Doyle, K. J. McDowall, M. J. Butler, and I. S. Hunter, "Characterization of an oxytetracycline-resistance gene, otrA, of Streptomyces rimosus," Molecular Microbiology, vol. 5, no. 12, pp. 2923–2933, 1991.

40. B. Halling-Sørensen, G. Sengeløv, and J. Tjørnelund, "Toxicity of tetracyclines and tetracycline degradation products to environmentally relevant bacteria, including selected tetracycline-resistant bacteria," Archives of Environmental Contamination and Toxicology, vol. 42, no. 3, pp. 263–271, 2002.

41. J. H. Kim, B. C. Campbell, N. Mahoney, K. L. Chan, R. J. Molyneux, and G. S. May, "Enhancement of fludioxonil fungicidal activity by disrupting cellular glutathione homeostasis with 2,5-dihydroxybenzoic acid," FEMS Microbiology Letters, vol. 270, no. 2, pp. 284–290, 2007.

42. H. J. Rosslenbroich and D. Stuebler, "Botrytis cinerea—history of chemical control and novel fungicides for its management," Crop Protection, vol. 19, pp. 557–561, 2000.

43. N. Ochiai, M. Fujimura, M. Oshima et al., "Effects of iprodione and fludioxonil on glycerol synthesis and hyphal development in Candida albicans," Bioscience, Biotechnology and Biochemistry, vol. 66, no. 10, pp. 2209–2215, 2002.

44. D. Hagiwara, Y. Asano, J. Marui, A. Yoshimi, T. Mizuno, and K. Abe, "Transcriptional profiling for Aspergillus nidulans HogA MAPK signaling pathway in response to fludioxonil and osmotic stress,"Fungal Genetics and Biology, vol. 46, no. 11, pp. 868–878, 2009.

45. T. Mizuno, "Two-component phosphorelay signal transduction systems in plants: from hormone responses to circadian rhythms," Bioscience, Biotechnology and Biochemistry, vol. 69, no. 12, pp. 2263–2276, 2005.

46. S. M. J. C. S. Cabral and J. P. S. Cabral, "Morphological and chemical alterations in Botrytis cinerea exposed to the dicarboximide fungicide vinclozolin," Canadian Journal of Microbiology, vol. 43, no. 6, pp. 552–560, 1997.

47. K. Fujii and S. Takamura, "Pyricut® (difulmetorim, UBF-002EC)—a new fungicide for ornamental use,"Agrochemicals Japan, no. 72, pp. 14–16, 1998.
48. "The E-pesticide manual," 2006.
49. J. Spiegel and G. Stammler, "Baseline sensitivity of Monilinia laxa and M. fructigena to pyraclostrobin and boscalid," Journal of Plant Diseases and Protection, vol. 113, no. 5, pp. 199–206, 2006.
50. B. N. Ziogas and S. G. Georgopoulos, "The effect of carboxin and of thenoyltrifluoroacetone on cyanide-sensitive and cyanide-resistant respiration of Ustilago maydis mitochondria," Pesticide Biochemistry and Physiology, vol. 11, no. 1–3, pp. 208–217, 1979.
51. M. Matsson and L. Hederstedt, "The carboxin-binding site on Paracoccus denitrificans succinate: quinone reductase identified by mutations," Journal of Bioenergetics and Biomembranes, vol. 33, no. 2, pp. 99–105, 2001.
52. K. Motoba, M. Uchida, and E. Tada, "Mode of antifungal action and selectivity of flutolanil," Agricultural and Biological Chemistry, vol. 52, no. 6, pp. 1445–1449, 1988.
53. S. R. M. De Bittencourt, J. O. M. Menten, C. A. Dos Santos Araki et al., "Eficiency of the fungicide carboxin + thiram in peanut seed treatment," Revista Brasileira de Sementes, vol. 29, no. 2, pp. 214–222, 2007.
54. D. L. Smith, M. C. Garrison, J. E. Hollowell, T. G. Isleib, and B. B. Shew, "Evaluation of application timing and efficacy of the fungicides fluazinam and boscalid for control of Sclerotinia blight of peanut,"Crop Protection, vol. 27, no. 3–5, pp. 823–833, 2008.
55. B. A. C. Ackrell, et al., "Structure and function of succinate dehydrogenase and fumarate reductase,"Chemistry and Biochemistry of Flavoenzymes, vol. 3, pp. 229–297, 1992.
56. A. N. Tucker and T. T. Lillich, "Effect of the systemic fungicide carboxin on electron transport function in membranes of Micrococcus denitrificans," Antimicrobial Agents and Chemotherapy, vol. 6, no. 5, pp. 572–578, 1974.
57. Z. Guo, H. Miyoshi, T. Komyoji, T. Haga, and T. Fujita, "Uncoupling activity of a newly developed fungicide, fluazinam [3-chloro-

N-(3-chloro-2,6-dinitro-4-trifluoromethylphenyl)-5-trifluoro methyl-2-pyridinamine]," Biochimica et Biophysica Acta, vol. 1056, no. 1, pp. 89–92, 1991.

58. U. Brandt, J. Schubert, P. Geck, and G. Von Jagow, "Uncoupling activity and physicochemical properties of derivatives of fluazinam," Biochimica et Biophysica Acta, vol. 1101, no. 1, pp. 41–47, 1992.
59. J. P. Seiler, "Toxicology and genetic effects of benzimidazole compounds," Mutation Research, vol. 32, no. 2, pp. 151–167, 1975.
60. N. E. McCarroll, A. Protzel, Y. Ioannou et al., "A survey of EPA/OPP and open literature on selected pesticide chemicals—III. Mutagenicity and carcinogenicity of benomyl and carbendazim," Mutation Research—Reviews in Mutation Research, vol. 512, no. 1, pp. 1–35, 2002.
61. K. Gupta, J. Bishop, A. Peck, J. Brown, L. Wilson, and D. Panda, "Antimitotic antifungal compound benomyl inhibits brain microtubule polymerization and dynamics and cancer cell proliferation at mitosis, by binding to a novel site in tubulin," Biochemistry, vol. 43, no. 21, pp. 6645–6655, 2004.
62. L. C. Davidse, "Benzimidazole fungicides: mechanism of action and biological impact," Annual Review of Phytopathology, vol. 24, pp. 43–65, 1986.
63. B. S. Koo, H. Park, S. Kalme et al., "α- and β-tubulin from Phytophthora capsici KACC 40483: molecular cloning, biochemical characterization, and antimicrotubule screening," Applied Microbiology and Biotechnology, vol. 82, no. 3, pp. 513–524, 2009.
64. K. Rathinasamy and D. Panda, "Suppression of microtubule dynamics by benomyl decreases tension across kinetochore pairs and induces apoptosis in cancer cells," FEBS Journal, vol. 273, no. 17, pp. 4114–4128, 2006.
65. M. J. Clément, K. Rathinasamy, E. Adjadj, F. Toma, P. A. Curmi, and D. Panda, "Benomyl and colchicine synergistically inhibit cell proliferation and mitosis: evidence of distinct binding sites for these agents in tubulin," Biochemistry, vol. 47, no. 49, pp. 13016–13025, 2008.

66. C. W. Bi, J. B. Qiu, M. G. Zhou, C. J. Chen, and J. X. Wang, "Effects of carbendazim on conidial germination and mitosis in germlings of fusarium graminearum and botrytis cinerea," International Journal of Pest Management, vol. 55, no. 2, pp. 157–163, 2009.
67. P. Sukul and M. Spiteller, "Metalaxyl: persistence, degradation, metabolism, and analytical methods,"Reviews of Environmental Contamination and Toxicology, vol. 164, pp. 1–26, 2000.
68. H. Buchenauer, Chemistry of Plant Protection. Part 6: Controlled Release, Biochemical Effects of Pesticides, Inhibition of Plant Pathogenic Fungi, Springer, New York, NY, USA, 1990.
69. S. Lee, C. D. Garzón, and G. W. Moorman, "Genetic structure and distribution of Pythium aphanidermatum populations in Pennsylvania greenhouses based on analysis of AFLP and SSR markers,"Mycologia, vol. 102, no. 4, pp. 774–784, 2010.
70. D. W. Holloman and K. Chamberlain, "Hydroxypyrimidine fungicides inhibit adenosine deaminase in barley powdery mildew," Pesticide Biochemistry and Physiology, vol. 16, no. 2, pp. 158–169, 1981.
71. J. K. M. Brown and C. G. Simpson, "Genetic analysis of DNA fingerprints and virulences in Erysiphe graminis f.Sp. hordei," Current Genetics, vol. 26, no. 2, pp. 172–178, 1994.

Chapter 6

Efficacy of Combined Formulations of Fungicides with Different Modes of Action in Controlling Botrytis Gray Mold Disease in Chickpea

M. H. Rashid[1], M. Ashraf Hossain[2], M. A. Kashem[3], Shiv Kumar[4], M. Y. Rafii[5,6], and M. A. Latif6,[7]

[1]RARS, Bangladesh Agricultural Research Institute, Rahmatpur, Barisal, Bangladesh

[2]Pulses Research Centre, BARI, Gazipur, Bangladesh

[3]Bangladesh Institute of Nuclear Agriculture, Mymensingh, Bangladesh

[4]International Centre for Agricultural Research in the Dry Areas, Aleppo, Syria

[5]Institute of Tropical Agriculture, Universiti Putra Malaysia, 43400 Serdang, Selangor, Malaysia

[6]Department of Crop Science, Faculty of Agriculture, Universiti Putra Malaysia, 43400 Serdang, Selangor, Malaysia

[7]Plant Pathology Division, Bangladesh Rice Research Institute, Gazipur 1701, Bangladesh

ABSTRACT

Botrytis gray mold (BGM) caused by Botrytis cinerea Pers. Ex. Fr. is an extremely devastating disease of chickpea (Cicer arietinum L.) and has a regional as well as an international perspective. Unfortunately, nonchemical methods for its control are weak and ineffective. In order to identify an effective control measure, six fungicides with different modes of action were evaluated on a BGM susceptible chickpea variety BARIchhola-1 at a high BGM incidence location (Madaripur) in Bangladesh for three years (2008, 2009, and 2010). Among the six fungicides tested, one was protectant [Vondozeb 42SC, a.i. mancozeb (0.2%)], two systemic [Bavistin 50WP, a.i. carbendazim (0.2%), and Protaf 250EC, propiconazole (0.05%)], and three combination formulations [Acrobat MZ690, dimethomorph 9% + mancozeb 60%, (0.2%); Secure 600WG, phenomadone + mancozeb (0.2%); and Companion, mancozeb 63% + carbendazim 12% (0.2%)]. The results showed superiority of combination formulations involving both protectant and systemic fungicides over the sole application of either fungicide separately. Among the combination fungicides, Companion was most effective, resulting in the lowest disease severity (3.33 score on 1–9 scale) and the highest increase (38%) of grain yield in chickpea. Therefore, this product could be preferred over the sole application of either solo protectant or systemic fungicides to reduce yield losses and avoid fungicide resistance.

INTRODUCTION

Globally, chickpea (Cicer arietinum L.) is the third most important pulse crop after dry beans and dry peas and is presently grown on 11.99 million ha with 10.94 million tons production [1]. Among the various biotic stresses of chickpea, botrytis gray mold (BGM) caused by Botrytis cinerea Pers. Ex. Fr. is the most devastating disease and has both regional and international perspective. It is an economically important disease in areas with cool, cloudy, and humid weather [2]. The crop encounters frequent BGM epidemics with near complete yield loss in the Indian Subcontinent accounting for 80% of the global chickpea area. Frequent BGM epidemics in north Indian states have been one of the reasons for a geographical shift of chickpea cultivation to southern states [3]. In Nepal, the recurring problem of BGM has caused a drastic reduction in chickpea area from 28,190 ha in 1990 to less than 9,000 ha in 2010. In Bangladesh, a conservative estimate of crop loss due to BGM ranges from 10 to 15% under normal conditions but can reach up to 80% or more under periods of high disease pressure [4, 5]. Therefore, chickpea cultivation has declined sharply from 102,867 ha in 1990 to 18,219 ha in 2000 [6] and further reduced to 7,224 ha in recent years. Yield losses due to BGM have also been reported from other chickpea growing countries including Australia, Argentina, Canada, Columbia, Hungary, Mexico, Myanmar, Spain, Turkey, the USA, and Vietnam [2, 7].

The fungus Botrytis cinerea is an opportunistic pathogen on a wide variety of crops, causing gray mold disease primarily through infections via wounds or dead plant parts. Various synthetic fungicides for controlling this disease have become ineffective due to the development of resistance [8, 9]. In chickpea, BGM is difficult to manage as the causal pathogen is soil, seed, and air borne and, unfortunately, nonchemical methods for its control are difficult and ineffective. Extensive screening of chickpea germplasm against BGM did not identify a genotype with a high level of resistance [2, 6, 10–16]. Therefore, integration of chemical application with cultural practices such as late sowing of erect cultivars at lower plant densities remains the only option to manage the disease commercially [5]. For chemical control, numerous fungicides with different modes of action are commercially available. Seed treatments with iprodione, mancozeb, thiabendazole, triadimefon, triadimenol, vinclozolin, thiram, benomyl, carbendazim, or cantan are effective in reducing seed infection [2, 5, 17–24]. Foliar sprays with captan, carbendazim, chlorothalonil, mancozeb, thiabendazole, thiram, triadimefon, or vinecloolin can offer some level of moderate control particularly when used in combination with seed dressing fungicides [5, 18]. All these fungicides belong to either protectant or systemic group of chemicals.

Traditionally, seed treatments with protectant fungicides have been utilized for controlling BGM of chickpea in Bangladesh. Protectant fungicides do not offer durable control against the disease since they are easily washed off the crop after rainfall. In Bangladesh, when rainfall is frequent, systemic fungicides often perform better than the protectant fungicides. It is reported that intense use of systemic fungicides especially benzimidazoles selects for resistant fungal strains of Botrytis cinerea across the crops [9, 25–27]. It is therefore recommended to use systemic fungicides responsibly for the control of BGM in chickpea. The use of formulated mixtures (coformulations) of two fungicides with different modes of action has often been recommended to help manage fungicide resistance development [28–30]. Recently, fungicides combining both protectant and systemic chemicals have been formulated and made commercially available to farmers mainly broaden the spectrum of activity to delay selection of resistant fungal populations and 8 to optimize efficacy [31]. The present study was, therefore, undertaken to assess the efficacy of coformulations in comparison to solo applications of protectant or systemic fungicides separately in controlling BGM disease in chickpea.

MATERIALS AND METHODS

Field experiments were conducted at the Pulse Research Substation of Bangladesh Agricultural Research Institute (Madaripur) Bangladesh in the postrainy season of three consecutive years (2007-08, 2008-09, and 2009-10). The experimental site is a known high incidence area for BGM disease in Bangladesh due to the environmental conditions prevalent during the crop season such as high relative humidity (>70%) and optimum air temperature (20–28°C) [4]. The experiment design was a randomized complete block with four replications. A BGM susceptible variety BARIchhola-1 was planted in rows spaced 40 cm apart in 12 m^2 (4 m × 3 m) plots. Six commercially available fungicides representing different modes of action, including systemic [Bavistin 50WP (carbendazim), BASF Bangladesh Ltd., and Protaf 250 EC (propiconazole), Auto Crop Care Ltd.)], protectant [Vondozeb 42 SC (mancozeb), Naafco Pvt. Ltd.], and combination of formulations [Acrobat MZ690 (dimethomorph 9% + mancozeb 60%), BASF Bangladesh Ltd.; Secure 600WG (phenomadone + mancozeb), Bayer

CropScience Ltd.; and Companion (mancozeb 63% + carbendazim 12%), Auto Crop Care Ltd.], and combination formulations [Acrobat MZ690 (dimethomorph 9% + mancozeb 60%), BASF Bangladesh Ltd.; Secure 600 WG (phenomadone + mancozeb), Bayer CropScience Ltd.; and Companion (mancozeb 63% + carbendazim 12%), Auto Crop Care Ltd.] were tested along with a nontreated control plot. These fungicides were applied as a foliar spray before the onset of the disease @ 0.2% except for Protaf 250 EC at 0.5%. First spray was done using a backpack low volume sprayer equipped with cone nozzle just before the onset of flowering stage followed by three subsequent sprays at 10-day intervals. Standard cultural practices typical to the area were achieved for weeding, insecticide, and fertility management. Insecticide applications were done as necessary. Insecticide Karate (0.2%) was applied to protect the crop from the infestation of Helicoverpa pod borer.

BGM occurred naturally and assessments were made three times: at 50% flowering stage, 10 days thereafter, and at 100% flowering stage. Disease assessment was based on a 1–9 scale [2]. In addition, assessments of plant height (cm), number of pods/plant, number of seeds/pod, and 100-seed weight (g) were recorded on 10 randomly selected plants from each plot. Grain yield (kg/ha) was estimated based on the yield harvested from the whole plot.

Data were evaluated using ANOVA and analyzed by MSTAT software with fungicide treatment in different years as fixed effects and blocks × blocks (year) as random effects. The nontreated control plot was used for comparing the relative performance of the treatments. Treatment means were separated by Fisher's Protected LSD at 0.05 probability level.

RESULTS AND DISCUSSION

ANOVA showed highly significant differences ($P<0.01$) among the treatments for disease severity, grain yield, and number of pods/plant in individual years. Pooled analysis showed highly significant differences among the fungicide treatments over the years. Nonsignificant effects of year and treatment × year interaction suggested that the efficacy of fungicide treatments was superior to nontreated control plots across the years for disease severity, grain yield, and number of pods/

plant. Therefore, data were summarized over the three-year period for comparison. For plant height, number of seeds/pod, and 100-seed weight, fungicide treatments were superior to nontreated control plots numerically but not statistically.

BGM Disease Intensity

Results of foliar spray of the fungicides on BGM in chickpea variety BARIchhola-1 showed significant reduction in disease severity with all the treatments compared to nontreated control plot (Table 1). The results on disease severity were similar over three years of experimentation and, therefore, data were averaged over the years for comparison. Coformulations involving both protectant and systemic fungicides showed superiority over the sole application of either systemic or protectant fungicides and, in turn, systemic fungicides were superior to protectant fungicides in controlling the disease (Figure 1). The average disease index was 3.67 with coformulations followed by 4.84 with systemic and 6.33 with protectant fungicides, compared to 7.33 in nontreated control plot. It is reported that while the protectant fungicides prevent only spore germination, the systemic fungicides inhibit fungal growth and sporulation [32]. Coformulations have the advantage of both not when the systemic partner is not active and, thus, their superiority was probably due to reduced infection at the onset of disease at early flowering stage and later on reduced growth and sporulation of the pathogen. Among the coformulations, foliar spray of Companion kept the disease minimum at 3.3 on a 1–9 scale (Table 1). Secure 600WG was the second most effective fungicide followed by Bavistin 50WP and Acrobat MZ in reducing disease severity. Earlier studies also reported superior efficacy of combined formulation of carbendazim + mancozeb in controlling collar and root rot diseases of strawberry and chilli caused by Sclerotium rolfsii [31, 33] and BGM disease of paprika caused by Botrytis cinerea [34]. In the present study, Companion showed additive effect, combining the advantage of both mancozeb and carbendazim. There have been many reports on the uses of mixtures of synthetic fungicides for the control of plant pathogenic fungi. When utilized in two-way mixtures, such fungicides maintain or enhance the level of control of a pathogen at reduced rates for both components utilized in combinations, compared to solo applied at higher rates [8, 35]. Therefore, coformulations should be preferred for effective control of BGM in chickpea.

Table 1: Effect of foliar spray of fungicides with different modes of action on BGM disease severity and grain yield in BGM susceptible chickpea variety "BARIchhola-1" under field conditions at Madaripur in Bangladesh during postrainy seasons of 2007-08, 2008-09, and 2009-10

Treatments	Disease severity (1–9 score)				Grain yield (kg/ha)				Yield increase over control (%)
	2007-08	**2008-09**	**2009-10**	**Mean**	**2007-08**	**2008-09**	**2009-10**	**Mean**	
Systemic fungicides									
Bavistin 50 WP	4	4	4	4.00	1610	1478	1705	1598	23.11
Protaf 250 EC	6	5	6	5.67	1552	1212	1651	1472	13.40
Protectant fungicides									
Vondozeb 42 SC	6	7	6	6.33	1402	1369	1793	1521	17.18
Combination formulation									
Acrobat MZ	4	4	4	4.00	1695	1713	1688	1699	30.89
Secure 600 WG	3	4	4	3.67	1703	1686	1910	1766	36.06
Companion	3	3	4	3.33	1785	1804	2174	1921	48.00
Mean	4	5	5	5	1625	1544	1820	1663	28
Control	8	7	7	7.33	1218	1139	1536	1298	—
CV (%)	—	—	—	—	12.35	12.35	7.94	—	—
LSD (0.05)	—	—	—	—	120.6	231.26	317.4	—	—

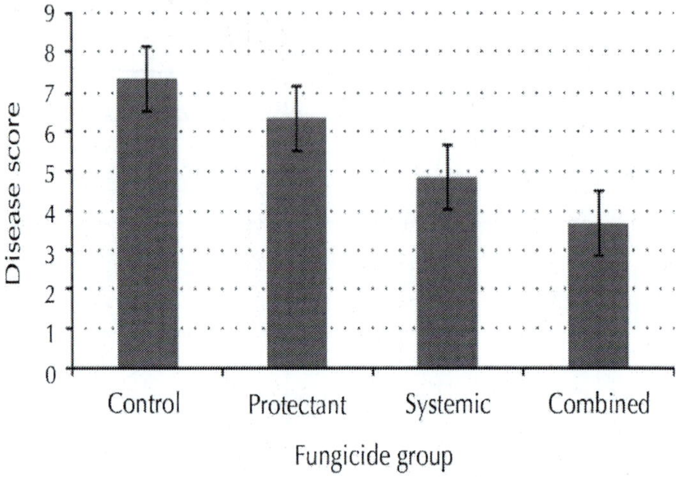

Figure 1: Effect of fungicides with different modes of action on Botrytis gray mold disease severity in susceptible chickpea variety "BARIchhola-1."

Grain Yield

Grain yield was significantly higher in all the fungicide treatment plots than in the nontreated control plot (Table 1). Yearwise and pooled analysis showed that coformulations had a significant yield advantage over the sole application of either systemic or protectant fungicides (Figure 2). Coformulations showed 38% increase in grain yield over the nontreated control plot, 18% over protectant, and 17% over systemic fungicides. Systemic and protectant fungicides were equally effective against BGM in terms of yield (17%) over the nontreated control plot. Companion was the most effective fungicide followed by Secure 600 WG, Acrobat MZ, and Bavistin 50 WP in increasing grain yield (Table 1). The highest average yield was achieved with Companion (1921 kg/ha) followed by Secure 600 WG (1766 kg/ha), Acrobat MZ (1699 kg/ha), and Bavistin 50WP (1598 kg/ha) compared to 1298 kg/ha in nontreated control plot. ANOVA revealed highly significant negative correlation ($r = -0.653^{**}$) between grain yield and BGM disease intensity in chickpea (Table 4), indicating that grain yield reduced sharply with increased disease intensity and fungicides are useful to harvest higher chickpea yield in BGM prone areas.

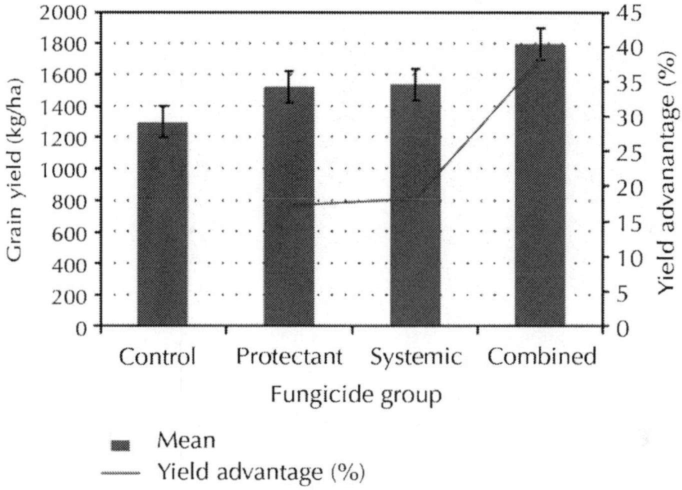

Figure 2: Effect of fungicides on grain yield of BGM susceptible chickpea variety "BARIchhola-1." Values are means of four replicates over three years.

Yield Components

The results presented in Tables 2 and 3 showed that combination formulations were superior to sole application of the fungicide components individually in terms of various yield components. However, their effect was statistically significant only for number of pods per plant and only numerically superior for number of seeds per pod, plant height, and 100-seed weight. The average number of pods per plant was highest in the Companion treated plots and lowest in the nontreated control plot. The average values for plant height, seeds/pod, and 100-seed weight were highest with foliar spray of Companion and lowest in the nontreated control plot. These results confirmed the earlier findings of [18, 36] that seed treatments with contact fungicides coupled with foliar spray with systemic fungicides provided better control over the individual control measures against BGM in chickpea. The results showed that the disease intensity had significantly negative correlations with number of pods per plant ($r = -0.730^{**}$) and seeds per pod ($r = -0.533^{**}$), suggesting that BGM in chickpea reduced grain yield mainly because of reduction in pods/plant and seeds/pod.

This is expected as the BGM disease initiates just before the onset of flowering, a critical stage which decides the plant capacity to bear the pods which, in turn, is highly positively correlated with grain yield ($r = 0.630^{**}$).

Table 2: Effect of foliar spray of fungicides with different modes of action on plant height and number of pods/plant in BGM susceptible chickpea variety "BARIchhola-1" under field conditions at Madaripur in Bangladesh during postrainy season of 2007-08, 2008-09, and 2009-10

Treatments	Plant height (cm)				Number of pods/plant			
	2007-08	2008-09	2009-10	Average	2007-08	2008-09	**2009-10**	**Average**
Bavistin 50 WP	54.00	46.77	52.50	51.09	37.70	42.17	36.73	38.87
Protaf 250 EC	52.40	45.93	51.63	49.99	36.23	34.27	33.33	34.61
Vondozeb 42 SC	55.10	46.77	56.20	52.69	35.47	46.57	38.60	40.21
Acrobat MZ	55.40	49.83	56.33	53.85	44.37	46.70	47.27	46.11
Secure 600 WG	54.80	46.20	56.43	52.48	41.70	47.17	45.40	44.76
Companion	55.60	48.60	57.60	53.93	53.57	53.57	47.47	51.54
Control	48.55	45.57	52.30	48.81	29.90	31.87	30.93	30.90
CV (%)	8.21	11.00	9.86	—	12.17	13.13	27.35	—
LSD (0.05)	7.33	9.75	9.63	—	5.48	6.16	18.35	—

Table 3: Effect of fungicices with modes of action on number of seeds/pod and 100-seed weight in BGM susceptible chickpea variety "BARI chhola-1" under field conditions at Madaripur in Bangladesh during the postrainy season of 2007-08, 2008-09, and 2009-10

Treatments	Number of seeds/pod				100-seed weight (g)			
	2007-08	2008-09	2009-10	Mean	2007-08	2008-09	2009-10	**Mean**
Bavistin 50 WP	1.52	1.55	1.37	1.48	12.17	11.83	15.67	13.22
Protaf 250 EC	1.57	1.53	1.50	1.53	11.93	10.27	15.00	12.40
Vondozeb 42 SC	1.55	1.53	1.30	1.46	11.43	12.33	13.00	12.25
Acrobat MZ	1.59	1.59	1.50	1.56	12.73	12.13	15.03	13.30
Secure 600 WG	1.55	1.60	1.53	1.56	12.43	12.00	15.00	13.14
Companion	1.59	1.63	1.51	1.58	13.00	12.43	15.00	13.48
Control	1.50	1.50	1.27	1.42	11.33	10.93	14.33	12.20
CV (%)	8.71	8.69	10.35	—	10.11	10.11	7.33	—
LSD (0.05)	0.18	0.15	0.29	—	2.05	0.65	1.92	—

Table 4: Correlation coefficient values of BGM disease intensity with grain yield and yield components in BGM susceptible chickpea variety "BARIchhola-1"

Trait	Grain yield	Pods/plant	Plant height	Seeds/pod	Seed weight
BGM score	−0.653**	−0.730**	−0.319	−0.533**	−0.236
Grain yield		0.630**	0.677**	0.062	0.681**
Pods/plant			0.198	0.524**	0.189
Plant height				−0.208	0.598**
Seeds/pod					−0.305

**Significant at 1% level; "r" values are based on observations recorded over three years.

CONCLUSIONS

Overall, the study clearly showed strong effect of fungicide applications on the reduction of disease severity and increase in grain yield and number of pods per plant. Coformulations showed additive effect of the protectant and systemic fungicides over solo application of each component fungicide in BGM management of chickpea. In the absence of host resistance, coformulations should be preferred over solo applications of systemic single-site fungicides to avoid resistance. Coformulations gave better disease control and higher grain yield in chickpea in BGM prone areas of South Asia than solo fungicides.

ACKNOWLEDGMENTS

The authors are grateful to the Ministry of Science and Technology (MST), Government of the People's Republic of Bangladesh, Dhaka, for supporting fund for successful completion of the study.

REFERENCES

1. ICRISAT, 2006, https://www.integratedbreeding.net/crop-information/chickpeas.
2. S. Pande, J. Galloway, P. M. Gaur et al., "Botrytis grey mould of chickpea: a review of biology, epidemiology, and disease management," Australian Journal of Agricultural Research, vol. 57, no. 11, pp. 1137–1150, 2006. · ·
3. M. Ali and S. Kumar, "Major technological advances in pulses: indian scenario," In Milestones in Food Legumes Research, M. Ali and S. Kumar, Eds., pp. 1–20, IIPR, Kanpur, India, 2009.
4. M. A. Bakr and F. Ahmed, "Botrytis gray mold of chickpea in Bangladesh," in Botrytis Gray Mold of Chickpea, M. P. Haware, D. G. Faris, and C. L. L. Gowda, Eds., pp. 10–12, ICRISAT, Patancheru, India, 1992.
5. C. Johansen, M. A. Bakr, M. S. Islam et al., "Integrated crop management of chickpea in environments of Bangladesh prone to Botrytis grey mould," Field Crops Research, vol. 108, no. 3, pp. 238–249, 2008. · ·

6. M. A. Bakr, S. A. Hussain, M. A. Afzal, and M. A. Rahman, "Chickpea status and production constraints in Bangladesh," in Integrated Management of Botrytis Gray Mold of Chickpea in Bangladesh and Australia, Summary Proceedings, Project Inception Workshop, pp. 19–32, BARI, Gazipur, Bangladesh, 2002.
7. Y. L. Nene, V. K. Sheila, and S. B. Sharma, A World List of Chickpea and Pigeonpea Pathogens, ICRISAT, Patancheru, India, 5th edition, 1996.
8. A. C. Pappas, "Evolution of fungicide resistance in Botrytis cinerea in protected crops in Greece,"Crop Protection, vol. 16, no. 3, pp. 257–263, 1997. · ·
9. P. Leroux, R. Fritz, D. Debieu et al., "Mechanisms of resistance to fungicides in field strains of Botrytis cinerea," Pest Management Science, vol. 58, no. 9, pp. 876–888, 2002. · ·
10. M. P. Haware and Y. L. Nene, "Screening chickpea for resistance to botrytis gray mold," International Chickpea Newsletter, vol. 6, pp. 17–18, 1982.
11. G. Singh, S. Kapoor, and K. Singh, "Screening of chickpea for gray mold resistance," International Chickpea Newsletter, vol. 7, pp. 13–14, 1982.
12. H. S. Chaubey, S. P. S. Beniwal, H. S. Tripathi, and Y. L. Nene, "Field screening of chickpea for resistance to Botrytis gray mold," International Chickpea Newsletter, vol. 8, pp. 20–21, 1983.
13. Y. P. S. Rathi, H. S. Tripathi, H. S. Chaubey, S. P. S. Beniwal, and Y. L. Nene, "Screening chickpea for resistance to botrytis gray mold," International Chickpea Newsletter, vol. 11, pp. 31–33, 1984.
14. Q. Ahmad, "Field screening of chickpea varieties and cultures against gray mold (Botrytis cinerea) in Bihar, India," International Chickpea Newsletter, vol. 21, p. 23, 1989.
15. B. B. Dewan, "Performance of chickpea lines in eastern Nepal," International Chickpea Newsletter, vol. 29, article 31, 1993.
16. H. S. Tripathi and Y. P. S. Rathi, "Resistance to Botrytis gray mould in chickpea: screening technique and identification of resistance sources," Journal of Mycology and Plant Pathology, vol. 30, pp. 231–232, 2000.

17. E. J. Cother, "Identification and control of root-rot fungi in Cicer arietinum (chickpea)," Plant Disease Reporter, vol. 61, pp. 736–740, 1977.
18. J. S. Grewal and S. K. Laha, "Chemical control of Botrytis blight of chickpea," Indian Phytopathology, vol. 36, pp. 516–520, 1983.
19. S. K. Laha and J. S. Grewal, "Botrytis blight of chickpea and its perpetuation through seed," Indian Phytopathology, vol. 36, pp. 630–634, 1983.
20. G. Singh and L. K. Bhan, "Chemical control of gray mold in chickpea," International Chickpea Newsletter, vol. 15, pp. 18–20, 1986.
21. G. Singh and L. Kaur, "Chemical control of gray mold of chickpea," Plant Disease Research, vol. 5, pp. 132–137, 1990.
22. M. A. Bakr, M. L. Rahman, F. Ahmed, and J. Kumar, "Progress in management of botrytis gray mold of chickpea in Bangladesh," in Recent Advances in Research in Botrytis Gray Mold of Chickpea, M. P. Haware, C. L. L. Gowda, and McDonald, Eds., pp. 17–19, ICRISAT, Patancheru, India, 1993.
23. M. P. Haware, "Diseases of chickpea," in The Pathology of Food and Pasture Legumes, D. J. Allen and J. M. Lenne, Eds., pp. 473–516, ICARDA, CAB International, Wallingford, UK, 1998.
24. S. Pande, G. Singh, J. N. Rao et al., Integrated Management of Botrytis Gray Mold of Chickpea, Information Bulletin No. 61, ICRISAT, Patancheru, India, 2002.
25. P. Leroux and M. Clerjeau, "Resistance of Botrytis cinerea Pers. and Plasmopara viticola (Berk. & Curt.) Berl. and de Toni to fungicides in French vineyards," Crop Protection, vol. 4, no. 2, pp. 137–160, 1985.
26. H. Nowacka, "Stability of resistance to benzimidazole fungicides in Venturia inaequalis," Review of Plant Pathology, vol. 7, pp. 691–696, 1992.
27. E. Stroe and A. A. Alexandri, "Tolerance of Venturia inaequalis (Cke.) Wint. of benzimidazole fungicides," Review of Plant Pathology, vol. 7, pp. 681–690, 1992.
28. C. J. Delp, Fungicide Resistance in North America, The American Phytopathological Society, St. Paul, Minn, USA, 1988.

29. Y. Elad, H. Yunis, and T. Katan, "Multiple fungicide resistance to benzimidazoles, dicarboximides and diethofencarb in field isolates of Botrytis cinerea in Israel," Plant Pathology, vol. 41, pp. 41–46, 1992.
30. M. Fujimura, "A new fungicide diethofencarb to cope with benzimidazole resistance," Japan Plant Protection, vol. 47, pp. 26–29, 1993.
31. S. Roy, A. Banerjee, J. Tarafdar, and S. K. Samanta, "Superior bio-efficacy of a combined formulation of carbendazim and mancozeb in inducing defense responses in chilli seedlings against Sclerotium rolfsii Sacc. in comparison with methyl jasmonate," Crop Protection, vol. 29, no. 2, pp. 163–167, 2010. · ·
32. V. Gisi, H. Binder, and E. Rimbach, "Synergistic interactions of fungicides with different modes of action," Transactions of the British Mycological Society, vol. 85, pp. 299–306, 1985.
33. H. Raj and S. D. Sharma, "Integrated management of collar and root rot (Sclerotium rolfsii) of strawberry," Acta Horticulturae (ISHS), vol. 696, pp. 375–379, 2005.
34. C. S. Yoon, E. H. Ju, Y. R. Yeoung, and B. S. Kim, "Survey of fungicide resistance for chemical control of Botrytis cinerea on paprika," Plant Pathology Journal, vol. 24, pp. 447–452, 2008.
35. R. E. Beever, E. P. Laracy, and H. A. Pak, "Strains of Botrytis cinerea resistant to dicarboximide and benzimidazole fungicides in New Zealand vineyards," Plant Pathology, vol. 38, pp. 427–437, 1989.
36. M. A. Bakr, M. L. Rahman, M. S. Hossain, A. U. Ahmed, and S. Pande, "Steps towards management of Botrytis gray mold of chickpea in Bangladesh," in Recent Advances in Research and Management of Botrytis Gray Mold of Chickpea, Summary Proceedings, Fourth Working Group Meeting to Discuss Collaborative Research on Botrytis Gray Mold of Chickpea, pp. 1523–1826, BARI, Gazipur, Bangladesh, 1998.

Chapter 7

Detection and Control of Ganoderma Boninense: Strategies and Perspectives

Roozbeh Hushiarian[1], Nor Azah Yusof[2,3], and Sabo Wada Dutse2,[4]

[1]Institute of Bioscience, Universiti Putra Malaysia, Serdang, Selangor 43400 UPM, Malaysia

[2]Department of Chemistry, Faculty of Science, Universiti Putra Malaysia, Serdang, Selangor 43400 UPM, Malaysia

[3]Institute of Advanced technology, Universiti Putra Malaysia, Serdang, Selangor 43400 UPM, Malaysia

[4]Department of Science Laboratory Technology, Hussaini Adamu Federal Polytechnic, Kazaure, Nigeria

ABSTRACT

The oil palm, an economically important tree, has been one of the world's major sources of edible oil and a significant precursor of biodiesel fuel. Unfortunately, it now faces the threat of a devastating disease. Many researchers have identified *Ganoderma boninense* as the major pathogen that affects the oil palm tree and eventually kills it. But identification of the pathogen is just the first step. No single method has yet been able to halt the continuing spread of the disease.

This paper discusses the modes of infection and transmission of *Ganoderma boninense* and suggests techniques for its early detection. Additionally, the paper proposes some possible ways of controlling the disease. Such measures, if implemented, could contribute significantly to the sustainability of the palm oil industry in South East Asia.

INTRODUCTION

Fungi which rot and eventually kill oil palm trees may be costing some South East Asian countries US$500 million a year. Indonesia and Malaysia jointly produce 84% of the world's total palm oil (Khairil and Hasmadi 2010; Ommelna et al. 2012). It has been reported that the economic loss caused by this pathogen is between RM225 million to RM1.5 billion (up to 500 million USD) a year (Ommelna et al. 2012; Arif et al. 2011).

The oil palm tree (*Elaeis guineensis* jacq.) originates from West Africa but was introduced to South East Asia by the British in the early 1870s. Oil palm trees start bearing fruit after 30 months of planting and are productive for 20 to 30 years. Needing just 0.26 hectares of land to produce one tone of oil, they are the most efficient oil-bearing crop in the world (Idris 2013). Unfortunately, infection with fungi has caused a decline in the productivity of oil palms and subsequently the palm oil industry, and created significant concern (Hartley 1967; Turner 1981).

A wide variety of fungi can affect and decay all types of woody plants and trees by colonizing in wounds on trunks, branches and roots. The disease developed is commonly called *wood rot disease* (Blanchette 1984) or *heartwood* when it is limited to the centre of older trees. Although the process may be very slow, it is estimated that the damage caused to timber trees by fungi, particularly in forests, is more than all other natural disasters combined, including insects and fire (Lonsdale et al. 2008).

Wood rot disease is classified according to each of the three areas affected - root rots, root and butt rots and stem rots (Arya and Perelló 2010). All wood rots grow inside the wood cells and degrade cell wall components. *Brown rot fungi* use cell wall polysaccharides and leave the lignin, while *white rot fungi* degrade lignin as well as all other wood components (Adaskaveg et al. 1993; Adaskaveg et al. 1991).

Although other fungi belonging to the Ascomycetes family can also decay wood, the species of fungi mostly responsible for developing complex wood rot diseases are Basidiomycetes and particularly the Ganoderma family (Khairuddin 1990; Rao 1990). Several different species of Ganoderma such as *Heterobasidion, Polyporus, Inonotus, Laetiporus, Phellinus, Chondrostereum,Peniophora, Lenzites, Pleurotus, Schizophyllum,* and *Trametes*are are responsible for a wide range of wood rots in a variety of trees (Seo and Kirk 2000; Moncalvo 2000). In all, at least seventy-five different examples of Ganoderma have been collected from twenty-one separate locations in Malaysia alone (Turner 1981). However, it is *Ganoderma boninense* which has been identified as the major disease of oil-palm trees (Khairuddin 1990; Rao 1990).

Ganoderma boninense (*G. boninsense*) causes both basal stem rot (BSR) and upper stem rot (USR) and remains South East Asia's most devastating oil palm diseases with direct loss of the stand, reduced yield of diseased palms and the resultant requirement for earlier replanting (Flood et al.2002). Once young palms show symptoms of the disease they usually die within 1 or 2 years, while mature trees can survive for only another 3 or so years (Corley and Tinker 2003).

Although it has been clearly identified as the main cause of the disease in oil palms, strategies for the early detection and control of *G. boninense* are still immature. The purpose of this paper is to outline the existing strategies, to evaluate their effectiveness and to suggest ways in which the spread of this destructive fungus might best be addressed.

Infection and Transmission

Considerable controversy remains but identifying the route of infection and the extent of pathogen diversity is critical to the development of effective disease control and plantation management (Cooper et al. 2011).

Ganoderma is characterized by basidiocarps large, perennial, woody brackets which are lignicolous and leathery, sometimes with a stem. The fruit bodies typically grow in a fan- or hoof-like form on the trunks of trees. They have double-walled, truncated spores with yellow to brown ornamented inner layers.

Delays in detection have been compounded by confusion between *G. zonatum,* weakly pathogenic to oil palm, and *G. boninense* the major pathogen (Pilotti 2005). High intraspecific variability found among *G. boninense* isolated from either neighbouring or distant trees supports the belief that sexual reproduction plays an important role in the epidemiology of Basal Stem Rot disease (Pilotti et al. 2003). Although there have been differences found among isolates from the same tree pointing to different strains (Miller et al. 1999), some researchers believe that the disease is a single taxon (Bridge et al. 2000).

Another reason the disease has not been well-detected in its early stages is because its cycle consists of a number of alternative and consecutive events: Firstly, there must be an injury to expose the wood. The cells around the injured area oxidize and discolour due to biochemical changes. This discolouration can develop further if the wound remains open to the many microorganisms which can land and grow on the moisture. The next most likely step is for bacteria and Ascomycetes or imperfect fungi to live on the wound, resulting in further discolouration, wetness of the area and erosion of parts of the cell wall. This is referred to as *wetwood, redheart*, or *blackheart*. Finally, as the wood-rotting fungi (Basidiomycetes) integrate, digestion of cell wall components begins. These microorganisms, however, are confined to the area which is already undergone the biochemical processes followed by bacterial reaction, Ascomycetes and imperfects. While these remain in tissues inside the endodermis in the initial stages their hyphae may be detected in different cells and tissues in advanced stages of disease. Infection of the stem leads to the formation of 'black lines' within the infected tissues (Ariffin et al. 1989 1991).

Ariffin and his colleagues had earlier revealed that the black line observed in the stem of oil palm infected with *G. boninense* is caused by a single mycelium and thus emphasizes the fungal origin of its formation (Ariffin et al. 1989). As presence of fungal hyphae almost exclusively on one side of the black line precludes the possibility of a dual infection involving another fungus and clearly indicates that *G. boninense* is the sole fungus present. Similarly, based on *in vitro*, morphological studies, *G. boninense* has been associated with the rotten upper stem. Flood and her colleagues stated that USR invariably causes discrete lesions originating from the frond base and spreading in successive wave of rot, each delimited by the brown or black line (Flood et al. 2002).

The ratio of USR to BSR in plantation estates ranges from 1:10 to 1:1 with some having an incidence of USR exceeding that of BSR (Hasan et al. 2005). In some commercial fields in Indonesia, the incidence of USR is higher than incidence of BSR (Flood et al. 2002) but its incidence in most Malaysian plantations is less than 5% (Turner and Gillbanks 1974).

Upper stem rot has symptoms similar to those of spear rot, bud rot, bunch rot and basal stem rot caused by root diseases. Lower leaves first become yellow and die from the tip to the base. This condition progresses to the middle of the crown, finally affecting the spear leaves. The stem tissues show a brown rot even when the roots of the palm are not affected (Hasan et al. 2005).

The symptoms of Basal stem rot are decay of the bottom of the stem from where basidiocarps emerge and sometimes also decay of the roots. Stem rotting restricts the uptake of water and nutrients to the fronds, causing chlorosis. When the disease is more advanced, the older fronds wilt and hang down to form a skirt around the trunk (Turner and Gillbanks 1974). Other observable symptoms are flattening of the crown and spear leaves that have not opened. In the most severe cases the stem might even fracture (Rees et al. 2012).

The whole process of discolouration and decay occurs within a few years to a few decades and is more common in older trees. Infection has been reported in four or five year old trees in replanted areas (Singh 1990) as well as in areas under plantation with coconut where *G. boninense* was found attacking seedlings and palms less than a year old in the nursery (Susanto 2009). The process does not necessarily complete and may stop at any stage due to quick healing of the wound, natural resistance of the wood or some antagonistic relationship between microorganisms. However, older trees are more likely to get injured during their lifespan and may be exposed to the disease process repeatedly.

Ganoderma spreads in the soil through roots and through the air. Compatibility studies have shown that fungi collected from the same field or area may have different origins so mycelial growth is probably not the only method of transmission of disease among the trees (Miller 1995). Although the development and spread of wood rots may differ from fungus to fungus and according to the type of tree, there are similarities. Basidiomycetes, like Ganoderma, have two strategies for reproduction: spores and mycelia. Other root-infecting basidiomycetes such as *Heterobasidi onannosum* spread from tree to tree through the soil by vegetative growth, often as one genet (Woodward et al. 1998) but studies of *G. boninense* have found that there is such diversity in oil palm plantations that infections by more than one genotype through sexual recombination with dispersal by spread of basidiospores has arisen (Pilotti et al. 2002).

Although they are certainly involved and there is evidence that they are able to germinate on cuts stems in plantations, there have been no successful infections of oil palms with basidiospores (Hasan et al. 2005; Idris 2013; Cooper et al. 2011). The reasons for this would seem to be *G. boninsense'* low aggressiveness and the need for large inocula (Rees et al. 2007).

Wind, rain and insects all assist to carry spores to wounds on trees, most commonly those which have been cut. In particular, the *Oryctes* beetle (Turner and Incorporated Society of Planters.1981) and larvae of the *Sufetula* spp caterpillar play at least a small role in the spreading of Ganoderma spores (Genty et al. 1976). Experiments in which an enormous number of Ganoderma spores were released in a field but did not infect most trees (Ho and Nawawi 1986) have indicated that infected tissues in the soil are more likely to spread the disease to healthy roots than airborne spores.

Thus, although not all scientists agree on how oil palms are infected and how the disease spreads, and indeed many studies including attempts to inoculate and infect oil palms with basidiospores have to date not been successful (Hasan and Flood 2003), there is, nevertheless, widespread acceptance that *G. boninsense* is the cause of both BSR and the less common USR (Navaratnam and Chee 1965; Lim et al. 1992; Sariah et al. 1994; Hasan and Turner 1998; Lim and Fong 2005; Breton et al. 2006).

Detection

Observation of such symptoms in the field as mature leaves wilting and falling through malnutrition or the presence of basidiomata of the pathogen on the tree was the only method of diagnosing disease in the early days (Lelong et al. 2010). The other early diagnostic method was drilling into diseased material in the tree for sampling and then using either a colorimetric method using ethlylenediaminetetraacetic acid (EDTA) (Natarajan et al. 1986) or a semi-selective media to cultivate

Ganoderma on agar plates (Darus et al. 1993). Both of these methods were time-consuming and not very accurate and it is now recognized that by the time basidiomata is apparent the disease is already well-established. Indeed there were no accurate techniques for detecting subclinical infections prior to the beginning of this century (Utomo and Niepold 2000).

The earliest molecular attempts were based on immunoassay. Antibodies were employed to detect Ganoderma in culture media (Reddy and Ananthanarayanan 1984; Darmono et al. 1993; Darmono and Suharyanto 1995). But many issues stood in the way of accurate and efficient techniques, including the lack of proper taxonomy information and confusion between some species of the genus (Moncalvo 2000; Ryvarden 1995; Paterson 2006a).

In an enzyme-linked immunosorbent assay, (ELISA), polyclonal antibodies were employed and showed relatively good negative results. Unfortunately, the test proved to be not species-specific as antibodies cross-reacted with other species (Utomo and Niepold 2000).

A study in which crude mycelium extract of *G. boninense* was used as an immunogene to generate monoclonal antibodies was found to give significantly better results than polyclonal antisera (Shamala et al. 2006) but further work is needed to produce monoclonal antibodies more specific to pathogenic species.

A promising Ganoderma-selective medium has facilitated sampling from the field for further studies (Ariffin et al. 2000). A more accurate alternative molecular diagnosis technique is to target oligonucleotides particularly DNA.

To design a reliable diagnostic molecular technique based on DNA sequencing, it is essential to have sufficient sequencing data. Table 1 shows that apart from manganese-superoxide dismutase, an enzyme related to the antioxidant defence mechanism of the cell, other available sequences are either related to ribosomal RNA (rRNA) or are designed molecular techniques for detection of the microorganism.

Table 1: Available nucleotide sequences of *Ganoderma boninense* in databases

Accession number	Sequence length (bp)	Product	Source
AAB16771	426	Manganese-superoxide dismutase	Strain RSH RS
U56128	683	Manganese-superoxide dismutase	Strain RSH RS
JQ665226	745	18S rDNA	Isolate G1
JQ665227	743	18S rDNA	Isolate G2
JQ665228	744	18S rDNA	Isolate G3
JQ665229	745	18S rDNA	Isolate G4
JQ665230	735	18S rDNA	Isolate G5
JQ665231	743	18S rDNA	Isolate G6
JQ665232	740	18S rDNA	Isolate G7
JQ665233	745	18S rDNA	Isolate G8
JQ665234	742	18S rDNA	Isolate G9
JQ665235	740	18S rDNA	Isolate G10
JQ665236	737	18S rDNA	Isolate G11
JQ665237	741	18S rDNA	Isolate G12
JQ665238	740	18S rDNA	Isolate G14
AF255198	1729	18S rDNA	Strain FA-PP28
FJ154775	884	18S rDNA	Isolate GR376
EU701010	297	18S rDNA, misc_RNA (ITS 1)	Strain FA5035
EU841913	297	18S rDNA, misc_RNA (ITS 1)	Strain FA5017
X78749	238	RS rDNA and ITS 1	Strain RS
X78770	207	RS rDNA and ITS 2	Strain RS
X78777	1458	25S rDNA	Strain RS
E38237	56	Method for detection	-
E38238	20	Method for detection	-
E38239	20	Method for detection	-
E38240	18	Method for detection	-

BD082757	616	Method for detection	-
BD082758	616	Method for detection	-
BD082759	612	Method for detection	-
BD082761	24	Method for detection	Synthetic construct
BD082762	24	Method for detection	Synthetic construct
BD082763	24	Method for detection	Synthetic construct
BD082764	23	Method for detection	Synthetic construct
BD082765	20	Method for detection	Synthetic construct
BD082766	22	Method for detection	Synthetic construct

Hushiarian *et al.*

Hushiarian *et al. SpringerPlus* 2013 **2**:555, doi:10.1186/2193-1801-2-555

The small ribosomal subunit RNA (in prokaryotes 16S and in eukaryotes 18S) gene is one of the most important molecular markers with a range of applications in biodiversity screening, phylogenetic analyses and evolutionary studies (Meyer et al. 2010). It is known to be a high similarity among sequences of rRNA (Kwon et al. 1991), particularly the 18S subunit and its secondary structure, within different species, and thus may be a reliable target for classification purposes (Neefs et al. 1990). A phylogenetic study of Ganoderma using single locus mt SSU rDNA led to dividing Ganoderma into six distinct monophyletic groups and indicated that complex situations related to the geographical region and the pathogen-host relationship must be considered as well as the phylogenetic relationships (Soon Gyu and Jung 2004). Nuclear 18S rDNA analysis may also be used to provide molecular evidence on the long distance dispersal of Ganoderma across the southern hemisphere (Moncalvo and Buchanan 2008) and has already shown the diversity of wood-decaying fungi in India (Singh et al. 2013).

A review on available data in the genebank following DNA sequences with accession numbers: BD082757, BD082758 and BD082759 which are retrieved from a patented method for detection

(Sakamoto et al. 2001) are aligned and shown to have little difference (Figure 1).

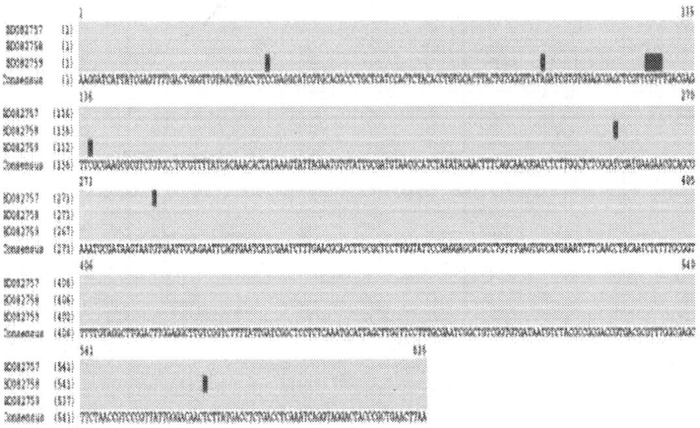

Figure 1: Multiple sequence alignment of part of 18S rDNA of *G. boninense* described in a patented method for detection.

A multiple sequence alignment of fragments of 18S rDNA from different isolates in Malaysia (Kok et al. 2013) shows close similarity in the region between bases 22 and 708 (Figure 2). Isolate G5 with almost 7% difference can be differentiated from the others.

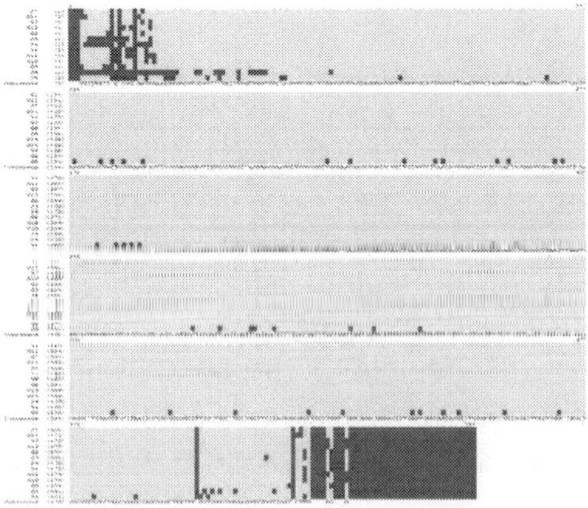

Figure 2: Multiple sequence alignment of partial 18S rDNA from isolates G1, G2, G3, G4, G5, G6, G7, G8, G9, G10, G11, G12, G14.

Sequences with accession numbers of EU701010 and EU841913 are 100% identical (Figure 3). They start with 45 bp of 18S rDNA.

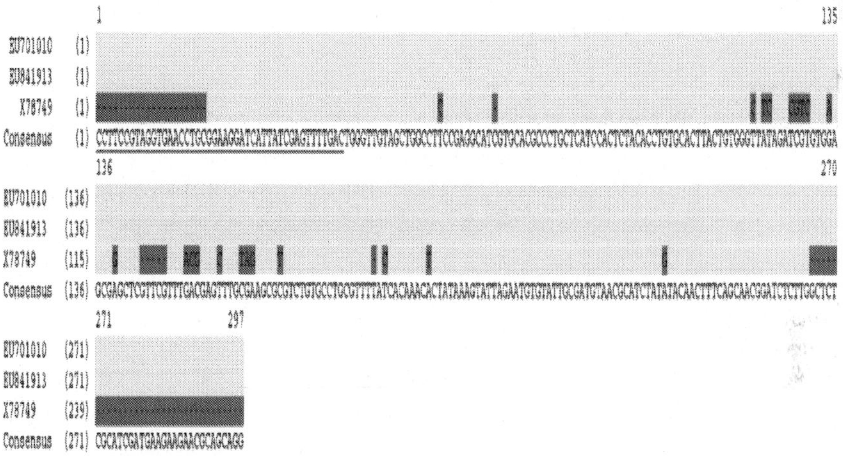

Figure 3: Sequence alignment between partial ITS1segments with accession numbers EU701010, EU841913 and X78749.

The alignment between the consensus sequences from Figures 1, 2 and 3 and the longest available sequence of 18S rRNA gene (1729 bp) with accession number AF255198 (Moncalvo and Buchanan2008) combined with other sequences in Table 1 produced a map of rDNA which is illustrated in Figure 4. This map is completed by comparison of these data with Moncalvo and Buchanan's findings and nearest relatives of *G. boninense*; *G. australe* and *G. pfeifferi* (Moncalvo et al. 1995). The D2 region in 25S rDNA is shown by the arrow. This region has been used for phylogenetic study and evolutionary relation between taxa and it has been shown that *G. boninense* and *G. pfeifferi* have just one nucleotide difference despite their 14 nucleotide difference in the Internal Transcribed Spacer (ITS) 1.

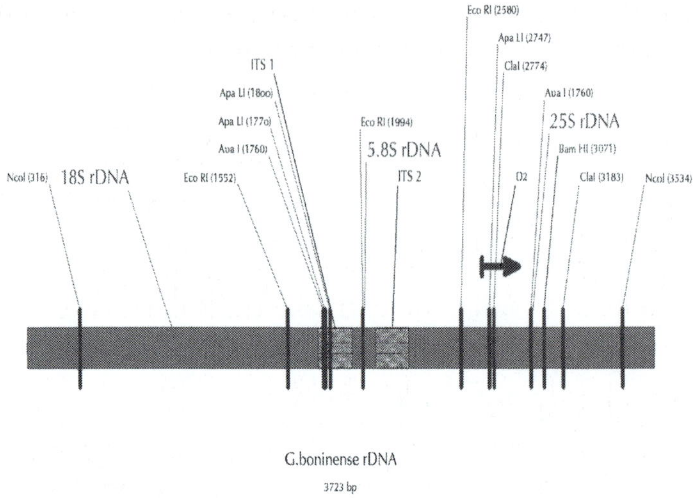

Figure 4: Schematic map of nuclear rDNA with related restriction endonuclease sites.

This information can be employed for the development of molecular techniques for detection purposes.

Polymerase Chain Reaction (PCR), by providing an opportunity for the amplification of certain regions of a template molecule (DNA or RNA), has been of great benefit to development and innovation in a variety of PCR-based techniques with diverse applications. PCR test using universal fungal primers for ITS region showed relatively more accuracy than ELISA (Utomo and Niepold2000).

Even now, it is still worthwhile for oil plantation farmers to use a combination of both procedures – the ELISA test for bulk screening and, where the results are positive, a subsequent PCR test to verify the results and to isolate Ganoderma (Utomo and Niepold 2000).

One of the most practical PCR-based techniques in characterization and detection of pathogenic fungi in plants is Random Amplification of Polymorphic DNA (RAPD). Here several arbitrary short primers are used with a long genomic DNA template in the expectation that some of them will amplify and provide a profile of the template (Williams et al. 1990; Williams et al. 1993). RAPD, together with microsatellite analysis, provides a further technique known as Random Amplification Microsatellite (RAMS) (Hantula et al. 1996). RAPD and RAMS can be adjusted (Zakaria et al. 2005) and employed to analyze different isolates of *Ganoderma* spp. (Zakaria et al. 2009).

Applying the Restriction Fragment Length Polymorphism (RFLP) technique in both highly conserved and variable sequences of ITS of rDNA at the species level provides an opportunity to facilitate genetic variation studies (Nusaibah et al. 2011). Since there is more divergence observed in ITS 1 rather than ITS 2 (Moncalvo and Buchanan 2008), it is recommended designing an experiment based on ITS 1 sequences. Furthermore, unlike ITS 2 there are 3 restriction sites in ITS 1 region. A PCR-RFLP method based on some sequential differences between pathogenic and non-pathogenic *Ganoderma* spp. has been designed and reported to be both more accurate and less sensitive to contamination than RAPD and Amplified fragment length polymorphism (AFLP) (Utomo et al. 2005) although the importance of internal amplification controls (IAC) is ignored in this method (Paterson 2007). Using IAC is highly recommended to avoid false negative results in PCR methods (Hoorfar et al. 2004).

The Malaysian Palm Oil Board (MPOB) has worked extensively on the early detection of Ganoderma. After an attempt to design a PCR technique (Idris et al. 2003), the Board worked on a polyclonal antibody for ELISA technique (Idris and Rafidah 2008) and finally provided a multiplex PCR-DNA kit (Idris et al. 2010). However, the PCR technique is vulnerable to contamination and since some secondary metabolites can inhibit PCR (Paterson 2004; Mizushina et al. 1999), purification of the DNA prior to the reaction is crucial. Regardless, the alkaline DNA extraction is simpler and quicker than other DNA extraction methods combined with purification procedures (Utomo and Niepold 2000).

FISH and DNA microarray are interesting options to the DNA-based molecular techniques outlined above. An electrochemical DNA biosensor has been designed and adjusted for detection of *G. boninense* (Dutse et al. 2012; Dutse et al. 2013) but as for all direct molecular techniques, the preparation of representative samples and extraction of DNA remains a challenge.

Meanwhile, as long as molecular methods remain complex and time-consuming, other techniques like E-nose, tomography and Hyperspectral reflectral data continue to be explored. A recent study successfully investigated the biomarkers (volatile compounds) of diseased oil palms for the development of specific Molecularly

Imprinted Polymer (MIP) sensors (Abdullah et al. 2012). This built on earlier studies which had used a commercially available electronic nose and an artificial neural network with 32 sensors for different trees. The odour profile recorded provided differentiating information that seems to be practical particularly if it is accompanied with chromatography techniques (Markom et al. 2009; Abdullah et al. 2012).

In another recent study, ergosterol, a membrane component of fungi, was investigated using high-performance liquid chromatography (HPLC). It is believed that quantification of this component accompanied by other techniques could provide a complex diagnosis system approaching the early detection aim (Mohd As'wad et al. 2011).

Another method to non-molecular approaches is a 2007 study undertaken by MohdShu'ud and his colleagues, who attempted to identify Ganoderma in oil plant stems from tomography images (Shu'ud et al. 2007). Expert rules were established in a Fuzzy Inference System and images from the trees triaged into 3 classifications: intact tissue, Ganoderma infection and infection other than Ganoderma. This model is obviously limited by the experts' knowledge of infection patterns in oil palms and to make it more feasible for widespread use, further testing to improve the rules is required.

Yet another non-molecular diagnostic method, claimed by its authors to be less complex and expensive and indisputably less damaging to the tree itself, is the use of statistical modeling to classify canopy spectra into Ganoderma attack severity levels (Lelong et al. 2010). First, field protocols and measurements are taken, a combination of analyses is applied and finally an algorithm is developed which has already been able to assess 4 levels of disease severity with 94% accuracy. Although this tool seems promising, especially in identifying seriously diseased trees from healthy ones, further work needs to be undertaken to more accurately discern the intervening levels of infection in the trees.

Overall, it has been and continues to be very difficult to find a species-specific detection method.

Control

The lack of sufficient data about *Ganoderma* spp. required for developing a reliable early stage diagnosis system inevitably leads to difficulty in control of the disease (Rolph et al. 2000). Currently, although it is impossible to manage a field free of pathogen (Sanderson et al. 2000), considerable reduction of the disease can be achieved through a proper management system of healthy stands to reduce the introduction of the pathogen to them. This is done by minimizing wounds on trees, by improving treatment and harvesting operations, and by clearing the old trees before extreme age susceptibility (Flood et al. 2000). It has been a routine practice to treat large wounds in other trees with paint or dressing and this seems just as applicable for oil palm trees, although its effectiveness in the prevention of decay is questionable. The use of fungicides tends to be another legitimate treatment for living stands, although this approach has yet to be studied and developed (Ariffin et al. 2000).

In fact, there are at least ten different techniques which have been used in attempts to manage the disease, with varying levels of success. Relatively ineffective measures include soil mounding, surgery and isolation trenching.

Soil Mounding

This method in which soil is heaped around the trunk to a height of 75 cm may prolongs the life of the tree but is not effective in controlling BSR (Lim et al. 1993; ChengTuck and Hashim 1997).

Surgery

Cutting out the dead tissue/basidiocarps by hand with a chisel (Turner 1981) or with a mechanical back-hoe (Singh 1990) has been tried with generally mediocre results, with the exception of some moderate success in small plantations (Cooper et al. 2011; Marshall et al. 2004).

More successful methods include isolation trenching, sanitation and leaving soil fallow after ploughing.

Isolating a diseased tree or stump with a 2×2 meter trench has been found to be a useful method of delaying BSR by up to 14 years because it stops contacts between roots (Hasan and Turner1998; Chung 2011).

Sanitation or Removal of Diseased Material

The removal of diseased materials is usually carried out in two situations – in existing plantings where BSR is evident, and at the replanting stage. For many years, burning the materials has been common but this is creating environmental issues in the region. Quarantining, shredding and isolating are other ways in which the roots of healthy palms can be prevented from coming into contact with those of diseased palms, stumps or roots (Singh 1990; Turner 1981; Khairuddin 1990; Idris et al. 2004; Chung 2011).

Ploughing and Harrowing

Large plantations often use 2 rounds of ploughing to 60cm depth and one round of harrowing to chop up the remaining roots in disease prone areas before planting new seedlings (Flood et al.2000).

Fallowing

Situations in which the soil has been left fallow in replanting systems have shown significant effect on later disease incidence. Further studies are continuing to assess the optimum length of fallow and the potential planting of other crops to balance the economic effect (Virdiana et al. 2010).

Planting Legume Cover Crops (LCC)

Mandatory in some plantations, the practice of planting ground covers, while controlling weeds and erosion, may in fact be introducing legume species which are themselves susceptible to *G. boninense* (Chung 2011) so must be undertaken with caution.

Chemical Treatments

Although investigated as an option, injection of fungicides such as hexaconazole is rarely practiced because this method has not shown to provide effective control (Chung 2011).

Fertiliser

It is still too soon to ascertain how effective fertilisers are against BSR. Experimentation with macro- and micro-elements such as K, N and P have resulted in some positive change to disease levels and plant productivity (Singh 1990; Chung 2011).

Biological Control

Antagonistic fungi have been under investigation for a few decades after successful protection of stumps against *Fomesannosus* (Rishbeth 1959 1963) followed by commercial control of the disease. From early studies on the biological control of wood rotting fungi, species of Trichoderma have attracted attention because of their potential antagonistic effect in controlling other fungi (Rifai 1969; Dennis and Webster 1971a, b, c). Trichoderma is considered to be able to control Ganoderma, successfully (Soepena et al. 2000). It has been shown that the disease is significantly lower in a field treated by a biological control agent (*T. harzianum*) than in an untreated field (Susanto et al. 2005). It has also been shown that production of certain fungal cell wall-degrading enzymes like glucanases and chitinases increase in oil palm trees in the presence of *Trichoderma spp.* (Ferreira et al. 2007) as a defence mechanism due to the host-pathogen interaction (López2010; Shrestha et al. 2008; Naher et al. 2011).

Filamentous fungi have chitin and β1,3-glucan in their hyphae so glucanases and chitinases have a synergistic effect on each other (Latgé 2007). The same defence reaction has previously been shown in tissues infected by *Ganoderma spp.* (Siswanto and Darmono 1998). Producing toxic secondary metabolites, however, raises serious concerns about fungal biocontrol by Trichoderma in oil palm fields (Paterson 2006b).

A wide range of endophytic bacteria from both gram positive and gram negative groups can help their host against plant pathogens through lysis or antibiotic activity. The mechanism through which these harmless bacteria induce systemic resistance (ISR) differs completely from systemic acquired resistance (SAR) in the signal transduction pathway (Kloepper and Ryu 2006). Different types of hosts from grass to woody plants take advantage of these endophytes. From the time that *Pseudomonas fluorescens* was shown to elicit ISR in cucumber against cucumber anthracnose (Wei et al. 1991) a diverse genera of bacteria mostly belonging to *Bacillus* and *Pseudomonas* have been found to elicit ISR in their hosts.

Antagonistic activity of endophytic bacteria against *G. boninense* has been studied intensively and *Burkholderiacepacia, Serratiamarcescens* and *Pseudomonas aeroginosa* have been introduced as candidates (Zaiton et al. 2006). In another attempt, the inhibitory effect of isolated endophytic bacteria from healthy oil palm roots on *G. boninense* was studied *in vitro*. Among 20 isolates, *P. aeroginosa* was the only identified bacteria to have a significantly strong inhibitory effect on mycelial growth in culture media (Bivi et al. 2010). In yet another approach, the effect of different isolates of *Bacillus* and Enterobacter was tested on *G. boninense* growth on oil palm seedlings (Suryanto et al. 2012).

Susceptibility differences among host populations can either be as a result of genetic resistance in the trees (Breton et al. 2006; Idris et al. 2004) or perhaps pathogenicity differences of isolates of *G. boninense* (Rees et al. 2007). It is suggested that breeding and selection of palms containing more lignin making them more resistant to the Ganoderma (Casler et al. 2002) may be another approach to the disease control and invites more investigation.

Meanwhile, the ability of trees to express glucanase and chitinase provides a potential target for genetically resistant trees. Inevitably, chitinases are used in growth and development mechanisms of tree (Santos et al. 2008). Several types of plant chitinases have been detected from different parts of various plants, and are involved in the defence mechanism against fungal pathogens (Liu et al. 2005; Ponstein et al. 1994). They have an acidic iso-electric point and accumulate in intercellular spaces (Metraux et al. 1989; Silipo et al. 2010) or a basic iso-electric point and remain in central vacuoles (Mauch and Staehelin 1989; Collinge et al. 1993). It is known that plants synthesize different isozymes of chitinase using a multigene family with diverse functionality (van Hengel et al. 2001; Zhong et al. 2002). Thus, the total chitinase activity does not necessarily relate to the host-pathogen interaction and recent studies are focussing on detection of the type of chitinase expressed as a result of fungal invasion (Yeoh et al. 2013; Naher et al. 2011) using cloning techniques and study of the gene expression. Similar studies have been implemented for glucanases (Yeoh et al. 2012).

Resistant Planting Materials and Screening for Resistance

In the longer term, the use of resistant planting materials offers the greatest hope for the future control of BSR. Many plantations are already screening their oil palm planting materials for resistance against Ganoderma (Idris et al. 2006; Chung 2011). Methods using infested wood block inoculum have been documented by Rees and colleagues who have also shown that shading of seedlings temperature both have a dramatic effect on disease (Rees et al. 2007).

Early molecular studies have not yet isolated defence-related genes - the genetics and nature of resistance by oil palms to Ganoderma is still unknown.

Ultimately, proper management of the field remains the most important part of disease control. Huge amounts of waste created by cutting down the old or dead stands were previously burned but burning in the field can cause further problems. Nutritious debris can be a favourable culture media for infectious fungi and potentially spread the disease (Flood et al. 2000) so it is recommended removing these from the site, especially before replantation (Flood et al. 2005; Panchal and Bridge 2005).

Conversely, an oil palm trunk enriched with all the required nutrients is considered recyclable for fertilizing the field, although a small proportion of nutrients are released despite the fast degradation. The decomposition pattern shows that higher parts of the trunk have less lignin than basal segments. Using specific methods to pulverize and spread debris to accelerate degradation, waste can be managed along with Ganoderma and other pests can be reduced but unfortunately this whole process tends to be uneconomic (Paterson 2007).

As early detection is still in under debate, it is not surprising that finding a practical, successful disease control system for BSR disease appears to be equally complicated, demanding a variety of techniques and strategies.

CONCLUSIONS

The white rot fungus, *Ganoderma boninense*, is now known as a major threat to the lucrative palm oil industry in south-east Asia. It is believed that inoculum left by the alternative host plants, the inoculum from infected trees spreading by mycelial root contact and airborne basidiospores are the three main ways this fungus spreads. Due to the shortage of data, there have been difficulties in designing an efficient rapid technique for reliable early detection but there have been some significant advances over simple observation.

Extensive work has been undertaken with PCR techniques and antibodies for the ELISA technique but the PCR technique has proven too vulnerable to contamination.

The next option, DNA-based nanosensors and DNA microarrays, have been found to be easier to operate, faster, more accurate and more economically viable than conventional PCR-based techniques. These need to be investigated further as the issue of DNA extraction and proper sampling remains a big challenge.

It is also possible that screening for secondary metabolites along with the components of fungi in the host could be an alternative technique to direct molecular methods.

In parallel, non-molecular methods such as E-nose, tomography and Hyperspectral reflective data are being investigated with some level of success but are still a long way off providing species-specific detection.

Overall, it would seem that the most promising research is molecular. DNA foot printing as well as genome sequencing is best placed to provide the data so critical for facilitating phylogenetic and systematic studies.

Until the ideal future solution of disease resistance becomes a reality, attempts to manage oil palm plantations will continue to be predominantly centred on BSR control by removing diseased material to prevent root contact from infected trees. This is done through isolation trenching, ploughing, harrowing, clearing, burning and fallowing before replanting the soil with seedlings. Sanitation remains the single most important measure at this time in the wait for Ganoderma resistant palms.

In conclusion, there is an urgent need for an intensive genome project is undertaken on *G. boninense*. This is a perfect example of a situation where the adage "prevention is better than cure" holds true. Fortunately, there has been some decrease in incidence of devastation since management of the fields has been better understood but a strong unmet demand for research into resistant oil palms and biological controls remains if the economic benefits of the palm oil industry are to be sustained in South East Asia.

REFERENCES

1. Abdullah AH, Adom AH, Shakaff AYM, Ahmad MN, Zakaria A, Saad FSA, Isa CMNC, Masnan MJ, Kamarudin LM (2012) Hand-Held Electronic Nose Sensor Selection System for Basal Stamp Rot (BSR) Disease Detection. In: Intelligent Systems, Modelling and Simulation (ISMS), 2012. : Third International Conference. pp 737-742, 10.1109/isms.2012.139
2. Adaskaveg JE, Blanchette RA, Gilbertson RL (1991) Decay of date palm wood by white-rot and brown-rot fungi. Can J Bot 69(3):615-629, 10.1139/b91-083
3. Adaskaveg JE, Miller RW, Gilbertson RL (1993) Wood decay, lignicolous fungi, and decline of peach trees in South Carolina. Plant Dis 77(7):707-711
4. Arif MS, Roslan A, Idris AS (2011) Economics of oil palm pests and Ganoderma disease and yield losses. In: Proceedings of the Third MPOB-IOPRI International Seminar: Integrated Oil Palm Pests and Diseases Management. Kuala Lumpur: Convention Centre.
5. Ariffin D, Idris AS, Hassan AH (1989) Significance of the black line within oil palm tissue decayed by Ganoderma boninense. Elaeis 1(1):11-16
6. Ariffin D, Idris AS, Hassan AH (1991) Histopathological studies on colonization of oil palm root by Ganoderma boninense. Elaeis 3(1):289-293
7. Ariffin D, Idris AS, Singh G (2000) Status of Ganoderma in oil palm. Wallingford: In. CABI. pp 49-68 doi:10.1079/9780851993881.0049

8. Arya A, Perelló AE (2010) Management of fungal plant pathogens. Cambridge, MA: CAB International.
9. Bivi MR, Farhana MS, Khairulmazmi A, Idris A (2010) Control of ganoderma boninense: a causal agent of basal stem rot disease in oil palm with endophyte bacteria *in vitro*. Int J Agric Biol 12(6):833-839
10. Blanchette RA (1984) Screening wood decayed by white rot fungi for preferential lignin degradation. Appl Environ Microbiol 48(3):647-653
11. Breton F, Hasan Y, Hariadi S, Lubis Z, De Franqueville H (2006) Characterization of parameters for the development of an early screening test for basal stem rot tolerance in oil palm progenies. J Oil Palm Res. 24-36 (Special issue, April 2006)
12. Bridge PD, O'Grady E, Pilotti CA, Sanderson FR (2000) Development of molecular diagnostics for the detection of Ganoderma isolates pathogenic to oil palm. In: Flood J, Bridge PD, Holderness M (eds) Ganoderma Diseases of Perennial Crops, Wallingford, UK: CABI Publishing. pp 225-247
13. Casler MD, Buxton DR, Vogel KP (2002) Genetic modification of lignin concentration affects fitness of perennial herbaceous plants. TAG Theoretical and applied genetics Theoretische und angewandte Genetik 104(1):127-131 Doi: 10.1007/s001220200015
14. ChengTuck H, Hashim K (1997) Usefulness of soil mounding treatments in prolonging productivity of prime-aged Ganoderma infected palms. Planter 73(854):239-244
15. Chung GF (2011) Management of ganoderma diseases in Oil palm plantations. Planter 87(1022):325-339
16. Collinge DB, Kragh KM, Mikkelsen JD, Nielsen KK, Rasmussen U, Vad K (1993) Plant chitinases. Plant J 3(1):31-40 Doi: 10.1046/j.1365-313X.1993.t01-1-00999.x
17. Cooper RM, Flood J, Rees R (2011) Ganoderma boninense in oil palm plantations: current thinking on epidemiology, resistance and pathology. Planter 87(1024):515-526
18. Corley R, Tinker P (2003) Vegetative propagation and biotechnology. The oil palm 4:201-215

19. Darmono TW, Suharyanto A (1995) Recognition of field materials of Ganoderma sp. associated with basal stem rot in oil palm by a polyclonal antibody. Menara Perkebunan 63(1):15-22
20. Darmono TW, Suharyanto DA, Moekti GR (1993) Polyclonal antibody against washing filtrate of mycelium culture of *Ganoderma* sp. Menara Perkebunan 61(3):67-72
21. Darus A, Seman I, Khairudin H (1993) Confirmation of Ganoderma infected palm by drilling technique. In: PORIM International Palm Oil Congress: Update and Vision (Agriculture). Kuala Lumpur, Malaysia: . pp 735-738
22. Dennis C, Webster J (1971) Antagonistic properties of species-groups of Trichoderma: I Production of non-volatile antibiotics. T Brit Mycol Soc 57(1):25 IN23. Doi: 10.1016/s0007-1536(71)80077-3
23. Dennis C, Webster J (1971) Antagonistic properties of species-groups of Trichoderma: II. Production of volatile antibiotics. T Brit Mycol Soc 57(1):41 IN44. Doi: 10.1016/s0007-1536(71)80078-5
24. Dennis C, Webster J (1971) Antagonistic properties of species-groups of Trichoderma: III. Hyphal interaction. T Brit Mycol Soc 57(3):363 IN362. Doi: 10.1016/s0007-1536(71)80050-5
25. Dutse SW, Yusof NA, Ahmad H, Hussein MZ, Zainal Z (2012) An electrochemical DNA biosensor for ganoderma boninense pathogen of the Oil palm utilizing a New ruthenium complex, [Ru (dppz) 2 (qtpy)] Cl2. Int J Electrochem Sci 7:8105-8115
26. Dutse SW, Yusof NA, Ahmad H, Hussein MZ, Hushiarian R (2013) DNA-based biosensor for detection of ganoderma boninense, an Oil palm pathogen utilizing newly synthesized ruthenium complex [Ru (phen) 2 (qtpy)] 2 based on a PEDOT-PSS/Ag nanoparticles modified electrode. Int J Electrochem Sci 8:11048-11057
27. Ferreira RB, Monteiro S, Freitas R, Santos CN, Chen Z, Batista LM, Duarte J, Borges A, Teixeira AR (2007) The role of plant defence proteins in fungal pathogenesis. Mol Plant Pathol 8(5):677-700 Doi: 10.1111/j.1364-3703.2007.00419.x
28. Flood J, Hasan Y, Turner PD, O'Grady EB (2000) The spread of Ganoderma from infective sources in the filed and its implications

for management of the disease in oil palm. In: Flood J, Bridge PD, Holderness M (eds) Ganoderma Diseases of Perennial Crops, Wallingford, UK: CABI Publishing. pp 101-112

29. Flood J, Hasan Y, Foster H (2002) Ganoderma diseases of oil palm-an interpretation from Bah Lias Research Station. Planter 78(921):689-710
30. Flood J, Keenan L, Wayne S, Hasan Y (2005) Studies on oil palm trunks as sources of infection in the field. Mycopathologia 159(1):101-107 Doi: 10.1007/s11046-004-4430-8
31. Genty P, De Chenon RD, Mariau D (1976) Infestation of the aerial roots of oil palms by caterpillars of the genus Sufetula (Lepidoptera:Pyralidae). Oleagineux 31(8/9):365-370
32. Hantula J, Dusabenyagasani M, Hamelin RC (1996) Random amplified microsatellites (RAMS) — a novel method for characterizing genetic variation within fungi. Eur J Forest Pathol 26(3):159-166 Doi: 10.1111/j.1439-0329.1996.tb00720.x
33. Hartley CWS (1967) The Oil Palm. Longmans. London: Green and Co. Ltd.
34. Hasan Y, Flood J (2003) Colonisation of rubber wood and oil palm blocks by monokaryons and dikaryons of Ganoderma boninense-implications to infection in the field. Planter 79(922):31-38
35. Hasan Y, Foster H, Flood J (2005) Investigations on the causes of upper stem rot (USR) on standing mature oil palms. Mycopathologia 159(1):109-112
36. Hasan Y, Turner P (1998) The comparative importance of different oil palm tissues as infection sources for basal stem rot in replantings. Planter 74(864):119-135
37. Ho YW, Nawawi A (1986) Isolation, growth and sporophore development of *Ganoderma boninense*from oil palm in Malaysia. Pertanika 9(1):69-73
38. Hoorfar J, Cook N, Malorny B, Wagner M, De Medici D, Abdulmawjood A, Fach P (2004) Letter to the Editor. J Appl Microbiol 96(2):221-222 Doi: 10.1046/j.1365-2672.2003.02188.x
39. Idris AS (2013) http://www.mpob.gov.my/ . Accessed 25 Sep 2013
40. Idris AS, Rafidah AR (2008) Polyclonal antibody for detection of *Ganoderma*. MPOB Inf Ser MPOB TT No. 405.

41. Idris AS, Yamaoka M, Hayakawa S, Basri MW, Noorhasimah I, Ariffin D (2003) PCR technique for detection of *Ganoderma*. MPOB Inf Ser MPOB TT No.188:MPOB TT No.188, MPOB TT No.188.
42. Idris A, Kushairi A, Ismail S, Ariffin D (2004) Selection for partial resistance in oil palm progenies to Ganoderma basal stem rot. J Oil Palm Res 16(2):12-18
43. Idris A, Kushairi D, Ariffin D, Basri M (2006) Technique for inoculation of oil palm germinated seeds with Ganoderma. MPOB Inf Ser 314:1-4
44. Idris AS, Rajinder S, Madihah AZ, Mohd BW (2010) Multiplex PCR-DNA kit for early detection and identification of *Ganoderma* species in oil palm. MPOB Inf Ser, MPOB TS No.73.
45. Khairil MMR, Hasmadi MI (2010) Applying GIS for mapping agriculture roads network in felda trolak utara for oil palm plantation management. Inform Manage Bus Rev 1(1):11-15
46. Khairuddin H (1990) Basal stem rot of oil palm: incidence, etiology and control. Selangor, Malaysia: Master of Agriculture Science thesis, Universiti Pertanian Malaysia, Selangor, Malaysia.
47. Kloepper J, Ryu C-M (2006) Bacterial Endophytes as Elicitors of Induced Systemic Resistance. In: Schulz BE, Boyle CC, Sieber T (eds) Microbial Root Endophytes, vol 9. Springer Berlin Heidelberg: Soil Biology, pp 33-52 doi:10.1007/3-540-33526-9_3
48. Kok SM, Goh YK, Tung HJ, Goh KJ, Wong WC, Goh YK (2013) *In vitro* growth of *Ganoderma boninense* isolates on novel palm extract medium and virulence on oil palm (*Elaeis guineensis*) seedlings. Mal J Microbiol 9(1):33-42
49. Kwon OY, Ogino K, Ishikawa H (1991) The longest 18S ribosomal RNA ever known. Eur J Biochem 202(3):827-833 Doi: 10.1111/j.1432-1033.1991.tb16439.x
50. Latgé J-P (2007) The cell wall: a carbohydrate armour for the fungal cell. Mol Microbiol 66(2):279-290 Doi: 10.1111/j.1365-2958.2007.05872.x
51. Lelong CC, Roger J-M, Brégand S, Dubertret F, Lanore M, Sitorus N, Raharjo D, Caliman J-P (2010) Evaluation of Oil-palm fungal

disease infestation with canopy hyperspectral reflectance data. Sensors 10(1):734-747

52. Lim K, Chuah J, Ho C Effects of soil heaping on Ganoderma infected oil palms. In: Proceedings of the (1993) PORIM International Palm Oil Congress. Update and Vision (Agriculture). 735-738
53. Lim H, Fong Y (2005) Research on basal stem rot (BSR) of ornamental palms caused by basidiospores from Ganoderma boninense. Mycopathologia 159(1):171-179
54. Lim T, Chung G, Ko W (1992) Basal stem rot of oil palm caused by Ganoderma boninense. Plant Pathol Bulletin 1(3):147-152
55. Liu J-J, Ekramoddoullah AKM, Zamani A (2005) A class IV chitinase is Up-regulated by fungal infection and abiotic stresses and associated with slow-canker-growth resistance to cronartium ribicola in western white pine (pinus monticola). Phytopathology 95(3):284-291 Doi: 10.1094/phyto-95-0284
56. Lonsdale D, Pautasso M, Holdenrieder O (2008) Wood-decaying fungi in the forest: conservation needs and management options. Eur J Forest Res 127(1):1-22 Doi: 10.1007/s10342-007-0182-6
57. López CM (2010) Nuevos aportes para la autenticidad del Popol Wuj. Revista Iberoamericana 75(226):125-152
58. Markom MA, Shakaff AYM, Adom AH, Ahmad MN, Hidayat W, Abdullah AH, Fikri NA (2009) Intelligent electronic nose system for basal stem rot disease detection. Comput Electron Agr 66(2):140-146 Doi: 10.1016/j.compag.2009.01.006
59. Marshall R, Hunt R, Pilotti C (2004) Low cost control for basal stem rot-a Poliamba initiative. Planter 80(936):173-176
60. Mauch F, Staehelin LA (1989) Functional implications of the subcellular localization of ethylene-Induced chitinase and [beta]-1, 3-glucanase in bean leaves. Plant Cell 1(4):447-457
61. Metraux JP, Burkhart W, Moyer M, Dincher S, Middlesteadt W, Williams S, Payne G, Carnes M, Ryals J (1989) Isolation of a complementary DNA encoding a chitinase with structural homology to a bifunctional lysozyme/chitinase. Proc Nat Acad Sci 3:896-900
62. Meyer A, Todt C, Mikkelsen NT, Lieb B (2010) Fast evolving 18S rRNA sequences from Solenogastres (Mollusca) resist standard

PCR amplification and give new insights into mollusk substitution rate heterogeneity. BMC Evol Biol 10:70 Doi: 10.1186/1471-2148-10-70

63. Miller RNG (1995) The characterization of Ganoderma population in oil palm cropping systems. UK: PhD thesis, University of Reading.

64. Miller RNG, Holderness M, Bridge PD, Chung GF, Zakaria MH (1999) Genetic diversity of Ganoderma in oil palm plantings. Plant Pathol 48(5):595-603 Doi: 10.1046/j.1365-3059.1999.00390.x

65. Mizushina Y, Takahashi N, Hanashima L, Koshino H, Esumi Y, Uzawa J, Sugawara F, Sakaguchi K (1999) Lucidenic acid O and lactone, new terpene inhibitors of eukaryotic DNA polymerases from a basidiomycete, ganoderma lucidum. Bioorg Med Chem 7(9):2047-2052 Doi: 10.1016/s0968-0896(99)00121-2

66. Mohd As'wad AW, Sariah M, Paterson RRM, Zainal Abidin MA, Lima N (2011) Ergosterol analyses of oil palm seedlings and plants infected with Ganoderma. Crop Prot 30(11):1438-1442 Doi: 10.1016/j.cropro.2011.07.004

67. Moncalvo JM (2000) Systematics of Ganoderma. In: Flood J, Bridge PD, Holderness M (eds) Ganoderma Diseases of Perennial Crops, Wallingford: CAB International. pp 23-45

68. Moncalvo J-M, Buchanan PK (2008) Molecular evidence for long distance dispersal across the southern hemisphere in the ganoderma applanatum-australe species complex (basidiomycota). Mycol Res 112(4):425-436 Doi: 10.1016/j.mycres.2007.12.001

69. Moncalvo J-M, Wang H-H, Hseu R-S (1995) Phylogenetic relationships in Ganoderma inferred from the internal transcribed spacers and 25S ribosomal DNA sequences. Mycologia 87(2):223-238

70. Naher L, Ho C-L, Tan SG, Yusuf UK, Abdullah F (2011) Cloning of transcripts encoding chitinases from Elaeis guineensis Jacq. and their expression profiles in response to fungal infections. Physiol Mol Plant Pathol 76(2):96-103 Doi: 10.1016/j.pmpp.2011.06.006

71. Natarajan S, Bhaskaran R, Shanmugan N (1986) Preliminary studies to develop techniques for early detection of Thanjavur wilt in coconut. Indian Coconut J 17(3):3-6

72. Navaratnam S, Chee K (1965) Root inoculation of oil palm seedlings with Ganoderma sp. Plant Dis 49:1011-1012
73. Neefs JM, Van de Peer Y, Hendriks L, De Wachter R (1990) Compilation of small ribosomal subunit RNA sequences. Nucleic Acids Res 18(Suppl):2237-2317
74. Nusaibah S, Latiffah Z, Hassaan A (2011) ITS-PCR-RFLP analysis of Ganoderma sp. Infecting industrial crops. Pertanika 34(1):83-91
75. Ommelna BG, Jennifer AN, Chong KP (2012) The potential of chitosan in suppressing Ganoderma boninense infection in oil-palm seedlings. J Sustain Sci Manage 7(2):186-192
76. Panchal G, Bridge PD (2005) Following basal stem rot in young oil palm plantings. Mycopathologia 159(1):123-127 Doi: 10.1007/s11046-004-4434-4
77. Paterson RR (2004) The isoepoxydon dehydrogenase gene of patulin biosynthesis in cultures and secondary metabolites as candidate PCR inhibitors. Mycol Res 108(12):1431-1437 Doi: 10.1017/s095375620400142x
78. Paterson RRM (2006) Fungi and fungal toxins as weapons. Mycol Res 110(9):1003-1010 Doi: 10.1016/j.mycres.2006.04.004
79. Paterson RRM (2006) Ganoderma – A therapeutic fungal biofactory. Phytochemistry 67(18):1985-2001 Doi: 10.1016/j.phytochem.2006.07.004
80. Paterson RRM (2007) Ganoderma disease of oil palm—A white rot perspective necessary for integrated control. Crop Prot 26(9):1369-1376 Doi: 10.1016/j.cropro.2006.11.009
81. Pilotti CA (2005) Stem rots of oil palm caused by <i>Ganoderma boninense: Pathogen biology and epidemiology. Mycopathologia 159(1):129-137 Doi: 10.1007/s11046-004-4435-3
82. Pilotti CA, Sanderson FR, Aitken EA (2002) Sexuality and interactions of monokaryotic and dikaryotic mycelia of Ganoderma boninense. Mycol Res 106(11):1315-1322
83. Pilotti CA, Sanderson FR, Aitken EAB (2003) Genetic structure of a population of Ganoderma boninense on oil palm. Plant Pathol 52(4):455-463 Doi: 10.1046/j.1365-3059.2003.00870.x
84. Ponstein AS, Bres-Vloemans SA, Sela-Buurlage MB, van den Elzen P, Melchers LS, Cornelissen B (1994) A novel pathogen-

and wound-inducible tobacco (nicotiana tabacum) protein with antifungal activity. Plant Physiol 104(1):109-118 Doi: 10.1104/pp. 104.1.109

85. Rao AK (1990) Basal stem rot (*Ganoderma*) in oil palm smallholdings. In: Ariffin D, Jalani S (eds) Proceedings of the Ganoderma Workshop, Palm Oil Research Institute of Malaysia, Bangi, Selangor, Malaysia: . pp 113-131
86. Reddy MK, Ananthanarayanan TV (1984) Detection of Ganoderma lucidum in betelnut by the fluorescent antibody technique. T Brit Mycol Soc 82(3):559-561
87. Rees RW, Flood J, Hasan Y, Cooper RM (2007) Effects of inoculum potential, shading and soil temperature on root infection of oil palm seedlings by the basal stem rot pathogen Ganoderma boninense. Plant Pathol 56(5):862-870 Doi: 10.1111/j.1365-3059.2007.01621.x
88. Rees R, Flood J, Hasan Y, Wills M, Cooper RM (2012) Ganoderma boninense basidiospores in oil palm plantations: evaluation of their possible role in stem rots of Elaeis guineensis. Plant Pathol 61(3):567-578
89. Rifai MA (1969) A revision of the genus Trichoderma. Kew: Mycol Papers 116 Commonwealth Mycological Institute.
90. Rishbeth J (1959) Stump protection against *Fomes annosus*: Treatment with substances other than creosote. Ann Appl Biol 47(3):529-541 Doi: 10.1111/j.1744-7348.1959.tb07285.x
91. Rishbeth J (1963) Stump protection against Fomes annosus. Ann Appl Biol 52(1):63-77 Doi: 10.1111/j.1744-7348.1963.tb03728.x
92. Rolph H, Wijesekara R, Lardner R, Abdullah F, Kirk PM, Holderness M, Bridge PD, Flood J (2000) Molecular variation in Ganoderma from oil palm, coconut and betelnut. In: Flood J, Bridge PD, Holderness M (eds) Ganoderma Diseases of Perennial Crops, Wallingford, UK: CABI Publishing. pp 205-221
93. Ryvarden L (1995) Can we trust morphology in Ganoderma?Buchanan PK, Hsea RS, Moncalvo JM (eds) Vancouver, Canada: Fifth International Mycological Congress. pp 19-24

94. Sakamoto A, Mikawa T, Murase M, Tanaka A, Kurane R (2001) Method for detecting *Ganoderma boninense* which is pathogenic fungus of *Elaeis guineensis* and primer for detection.
95. Sanderson FR, Pilotti CA, Bridge PD (2000) Basidiospores: their influence on our thinking regarding a control strategy for basal stem rot. In: Flood J, Bridge PD, Holderness M (eds) Ganoderma Diseases of Perennial Crops, Wallingford, UK: CABI Publishing. pp 113-119
96. Santos P, Fortunato A, Ribeiro A, Pawlowski K (2008) Chitinases in root nodules. Plant Biotechnol 25(3):299-307
97. Sariah M, Hussin M, Miller R, Holderness M (1994) Pathogenicity of Ganoderma boninense tested by inoculation of oil palm seedlings. Plant pathol 43(3):507-510
98. Seo GS, Kirk PM (2000) Ganodermataceae: nomenclature and classification. Wallingford: In. CABI. pp 3-22 doi:10.1079/9780851993881.0003
99. Shamala S, Chris D, Sioban O, Idris A (2006) Preliminary studies on the development of monoclonal antibodies against mycelia of Ganoderma boninense, the causal pathogen of basal stem rot of oil palm. Mal J Microbiol 2(1):30-34
100. Shrestha CL, Oña I, Muthukrishnan S, Mew TW (2008) Chitinase levels in rice cultivars correlate with resistance to the sheath blight pathogen Rhizoctonia solani. Eur J Plant Pathol 120(1):69-77 Doi: 10.1007/s10658-007-9199-4
101. Shu'ud MM, Loonis P, Seman IA (2007) Towards automatic recognition and grading of ganoderma infection pattern using fuzzy systems. T Eng Comput Technol 19:
102. Silipo A, Erbs G, Shinya T, Dow JM, Parrilli M, Lanzetta R, Shibuya N, Newman M A, Molinaro A (2010) Glyco-conjugates as elicitors or suppressors of plant innate immunity. Glycobiology 20(4):406-419 Doi: 10.1093/glycob/cwp201
103. Singh G (1990) The scourge of oil palms in the coastal areas. Planter 67:421-424
104. Singh SK, Doshi A, Pancholy A, Pathak R (2013) Biodiversity in wood-decay macro-fungi associated with declining arid zone trees of India as revealed by nuclear rDNA analysis. Eur J Plant Pathol. 1-10 10.1007/s10658-013-0172-0

105. Siswanto , Darmono TW (1998) Chitinase and β-1,3- glucanase activities against Ganoderma sp. in oil palm. Bogor: Biotechnology Research Unit for Estate Crops. pp 104-114

106. Soepena H, Purba RY, Pawirosukarto S (2000) A control strategy for basal stem rot (*Ganoderma*) on oil palm. In: Flood J, Bridge PD, Holderness M (eds) Ganoderma Diseases of Perennial Crops, Wallingford, UK: CABI Publishing. pp 83-88

107. Soon Gyu H, Jung HS (2004) Phylogenetic analysis of ganoderma based on nearly complete mitochondrial small-subunit ribosomal DNA sequences. Mycologia 96(4):742-755

108. Suryanto D, Wibowo RH, Siregar EBM, Munir E (2012) A possibility of chitinolytic bacteria utilization to control basal stems disease caused by Ganoderma boninense in oil palm seedling. Afr J Microbiol Res 6(9):2053-2059

109. Susanto A (2009) Basal stem rot in Indonesia. Biology, economic importance, epidemiology, detection and control. In: Proceedings of International Workshop on Awareness, Detection and Control of Oil Palm Devastating Diseases. Kuala Lumpur Convention Centre. Malaysia: Universiti Putra Malaysia Press.

110. Susanto A, Sudharto PS, Purba RY (2005) Enhancing biological control of basal stem rot disease (<i>Ganoderma boninense) in oil palm plantations. Mycopathologia 159(1):153-157 Doi: 10.1007/s11046-004-4438-0

111. Turner PD (1981) Oil palm diseases and disorders. Kuala Lumpur: Oxford University Press.

112. Turner PD, Gillbanks R (1974) Oil palm cultivation and management. Kuala Lumpur: Incorporated Society of Planters. p 672

113. Turner PD, Incorporated Society of Planters (1981) Oil palm diseases and disorders. Kuala Lumpur: Published for the Incorporated Society of Planters by Oxford University Press.

114. Utomo C, Niepold F (2000) Development of diagnostic methods for detecting ganoderma–infected Oil palms. J Phytopathol 148(9–10):507-514

115. Utomo C, Werner S, Niepold F, Deising HB (2005) Identification of Ganoderma, the causal agent of basal stem rot disease in oil palm using a molecular method. Mycopathologia 159(1):159-170 Doi: 10.1007/s11046-004-4439-z

116. van Hengel AJ, Tadesse Z, Immerzeel P, Schols H, van Kammen A, de Vries SC (2001) N-acetylglucosamine and glucosamine-containing arabinogalactan proteins control somatic embryogenesis. Plant Physiol 125(4):1880-1890 Doi: 10.1104/pp. 125.4.1880

117. Virdiana I, Hasan Y, Aditya R, Flood J (2010) Testing the effects of oil palm replanting practices (windrowing, fallowing and poisoning) on incidence of Ganoderma. Proc IOPC.

118. Wei G, Kloepper JW, Tuzun S (1991) Induction of systemic resistance of cucumber to Colletotrichum orbiculare by select strains of plant growth-promoting rhizobacteria. Phytopathology 81(11):1508-1512

119. Williams JG, Kubelik AR, Livak KJ, Rafalski JA, Tingey SV (1990) DNA polymorphisms amplified by arbitrary primers are useful as genetic markers. Nucleic Acids Res 18(22):6531-6535

120. Williams JGK, Hanafey MK, Antoni Rafalski J, Tingey SV (1993) [51] Genetic analysis using random amplified polymorphic DNA markers. In: Ray W (ed) Methods in Enzymology, vol Volume 218, 704, pp: Academic. pp 704-740 doi:http://dx.doi.org/10.1016/0076-6879(93)18053-F

121. Woodward S, Stenlid J, Karjalainen R, Hüttermann A (1998) Heterobasidion annosum: biology, ecology, impact and control. Cab International.

122. Yeoh KA, Othman A, Meon S, Abdullah F, Ho CL (2012) Sequence analysis and gene expression of putative exo- and endo-glucanases from oil palm (Elaeis guineensis) during fungal infection. J Plant Physiol 169(15):1565-1570 Doi: 10.1016/j.jplph.2012.07.006

123. Yeoh KA, Othman A, Meon S, Abdullah F, Ho CL (2013) Sequence analysis and gene expression of putative oil palm chitinase and chitinase-like proteins in response to colonization of Ganoderma boninense and Trichoderma harzianum. Mol Biol Rep 40(1):147-158 Doi: 10.1007/s11033-012-2043-8

124. Zaiton S, Sariah M, Zainal Abidin MA (2006) Isolation and characterization of microbial endophytes from oil palm roots: implication as biological control agents against Ganoderma. Planter 82(966):587-597

125. Zakaria L, Kulaveraasingham H, Guan TS, Abdullah F, Wan HY (2005) Random amplified polymorphic DNA (RAPD) and random amplified microsatellite (RAMS) of Ganoderma from infected oil palm and coconut stumps in Malaysia. Asia Pac J Mol Biol Biotechnol 13:23-34

126. Zakaria L, Ali N, Salleh B, Zakaria M (2009) Molecular analysis of Ganoderma species from different hosts in peninsula Malaysia. J Biol Sci 9(1):12-20

127. Zhong R, Kays SJ, Schroeder BP, Ye Z-H (2002) Mutation of a chitinase-like gene causes ectopic deposition of lignin, aberrant cell shapes, and overproduction of ethylene. Plant Cell 14(1):165-179 Doi: 10.1105/tpc.010278

Chapter 8

Comparative Assessment of Plant Protection Products: How many Cases will Regulatory Authorities have to Answer?

Michael Faust[1], Carolina Vogs[2], Stefanie Rotter[2], Janina Wöltjen[3], Andreas Höllrigl-Rosta[3], Thomas Backhaus[1], and Rolf Altenburger[2]

[1]Faust & Backhaus Environmental Consulting, Fahrenheitstr. 1, Bremen 28359, Germany

[2]Department Bioanalytical Ecotoxicology, UFZ - Helmholtz Centre for Environmental Research, Permoser Str. 15, Leipzig 04318, Germany

[3]Federal Environment Agency, Wörlitzer Platz 1, Dessau-Roßlau 06844, Germany

ABSTRACT

Background

The substitution principle has been included in the EU pesticides legislation as a new element. Comparative assessments will have to be conducted for all uses of plant protection products (PPPs) that contain active substances with certain hazardous properties, the so-called candidates for substitution (CFS). This study investigated the resulting workload in terms of the number of cases for comparative assessments that regulatory authorities may have to face. The analysis was carried out for Germany as an example.

Main Text

In Germany, the requirement for comparative assessments may affect up to 25% of all PPPs and around 50% of all uses of PPPs. In absolute terms, these are around 350 candidate products with 1,850 different uses. Alternative products without CFS may be available for around 40% of these uses. On average, a candidate product is authorised for around 18 different uses. For 11 of these uses, no alternatives are authorised. For the remaining seven uses, slightly more than seven alternatives are available on average. Multiplication of these factors gives an indicative figure of around 18,500 possible pairwise comparisons of candidate products with alternative products for every common use.

Conclusions

The high number of expectable cases poses a formidable challenge for the efficient conduct of the new task of comparative assessments by competent Member States authorities. To this end, new data handling systems, assessment procedures, and decision rules need to be established.

BACKGROUND

The Substitution Principle in the EU Pesticides Legislation

The substitution principle is a new element of the legislation on plant protection products (PPPs) in the European Union (EU). It was introduced with the new Regulation (EC) No 1107/2009 [[1]], in the following shortly denoted as the PPP Regulation. This replaced the old Directive 91/414/EEC on PPPs [[2]] in June 2011. In parallel, the substitution principle was also included in the new Regulation (EU) No 528/2012 on biocidal products [[3]], which came into force in September 2013. PPPs and biocidal products are collectively denoted as 'pesticides' under EU law, as has been defined in Article 3 of Directive 2009/128/EC on the sustainable use of pesticides in the European Community (EC) [[4]]. As a common rule, pesticides shall not be placed on the market or used unless they have been authorised in accordance with the applicable regulations. In general, 'substitution' of pesticides means that an authorisation is refused or withdrawn in favour of an alternative product or a non-chemical control or prevention method which presents a 'significantly lower risk', according to Annex IV of the PPP Regulation (EC) No 1107/2009 and Article 23 of the biocidal products Regulation (EU) No 528/2012, respectively. In detail, the conditions, rules, and criteria for applying the substitution principle differ for PPPs and for biocidal products. In this paper, we focus on substitution under the Regulation for PPPs.

The inclusion of the substitution aspect in the EU pesticides legislation is an outcome of a broader and long-lasting discussion about the guiding principles of chemicals regulation under EU law. As a generic policy principle, substitution means the replacement of hazardous chemical substances and products by less hazardous alternatives [[5]]. Whether this idea should be established as a legal demand for actors in the field has been subject to heated debates. Opponents, such as the German chemical industries for instance, argued that substitution was superfluous if safe use of a hazardous chemical could be ensured by appropriate risk management measures [[6]]. In 2001, during the preparation of the REACH legislation, the

Commission of the European Communities (COM) considered the substitution of hazardous chemicals as one of the 'key elements' of the proposed 'Strategy for a future Chemicals Policy' [[7]]. Five years later, in the final REACH legislation [[8]], legal requirements for feasibility analyses for substitution were, however, confined to substances of very high concern (SVHC) that are subject to authorisation (Article 55 of Regulation (EC) No 1907/2006). In all three pieces of legislation, where substitution has now been included as an element of authorisation procedures (REACH, biocidal products and PPPs), hazardous properties of chemicals serve only as a trigger for considerations for substitution, but are considered insufficient for decision making. Instead, comparative risk assessments of products have to be conducted as the basis for substitution decisions, which is novel and challenging.

Conventional risk assessments for individual PPPs, as they have been established under the old Directive 91/414/EEC, aim to ensure that regulatory acceptable exposure levels are not exceeded, but they do not provide incentives for reducing risks any further. This is changed by the complementary instrument of comparative risk assessment which supports a process of continuous improvement by identifying those PPPs that allow to achieve a desired purpose with minimal risks at a given point in time. This is particularly favourable for environmental risks, where the authorisation requirements still allow to tolerate temporary adverse effects as acceptable. Moreover, acceptable exposure levels for many pesticides on the market are only achievable by applying risk mitigation measures, such as protective equipment for workers or buffer zones between sprayed agricultural land and surface waters. Such measures may fail accidentally or may be disregarded negligently. Substitution of such products by alternatives that require less risk mitigation measures is therefore desirable and shall be supported by the new instrument of comparative assessments.

While the intended improvements are clear, the detailed procedures and methodologies for applying the substitution principle are not. Only in the Nordic countries, particularly in Sweden, the principle has been included in the national chemicals legislation since the beginning of the 1990s [[9]]. Other EU Member States (MS) have no comparable legislative tradition. Against this background, there is high uncertainty about potential impacts of this new element of EU pesticides legislation and the best way towards its efficient implementation.

Candidates for Substitution

Plant protection products contain one or more active substances. Under EU law, PPPs are authorised on the Member States level, while active substances are approved on the Community level. Approved active substances are included in a positive list established by the European Commission. Member States shall not authorise PPPs that contain active substances other than those on the positive list. Authorisations are only granted for specified uses, usually defined by a combination of a protected crop and a targeted pest.

The revised legislation now requires that certain active substances shall be approved by the European Commission only as 'candidates for substitution' (CFS) and listed separately from other approved active substances. Member States shall not grant authorisation to PPPs that contain such CFS, if a comparative assessment reveals that a significantly safer alternative is available for the same use.

CFS are active substances that have one or more of the hazardous properties listed in Table 1. As laid down in the PPP Regulation, their identification constitutes one task within the regular assessment of active substances on Community level. In order to speed up the process for the already approved active substances, an obligation for the European Commission (COM) was included in the PPP Regulation to establish an initial list of CFS until the end of 2013. However, completion of this task is now expected to be delayed by a few months [[10]]. At the time of writing of this manuscript (March 2014), the official list was not yet available.

Table 1: Criteria for the identification of active substances as candidates for substitution (CFS)

Number	**Legal text (Regulation (EC) No 1107/2009, Annex II, point 4)[[1]]a**
1	Its ADI, ARfD, or AOEL is significantly lower than those of the majority of the approved active substances within groups of substances/use categories
2	It meets two of the criteria to be considered as a PBT substance
3	There are reasons for concern linked to the nature of the critical effects (such as developmental neurotoxic or immunotoxic effects) which, in combination with the use/exposure patterns, amount to situations of use that could still cause concern, for example, high potential of risk to groundwater; even with very restrictive risk management measures (such as extensive personal protective equipment or very large buffer zones)
4	It contains a significant proportion of non-active isomers
5	It is or is to be classified, in accordance with the provisions of Regulation (EC) No 1272/2008, as carcinogen category 1A or 1B, if the substance has not been excluded in accordance with the criteria laid down in point 3.6.3[b]
6	It is or is to be classified, in accordance with the provisions of Regulation (EC) No 1272/2008, as toxic for reproduction category 1A or 1B if the substance has not been excluded in accordance with the criteria laid down in point 3.6.4[b]
7	If, on the basis of the assessment of Community or internationally agreed test guidelines or other available data and information, reviewed by the Authority, it is considered to have endocrine disrupting properties that may cause adverse effects in humans if the substance has not been excluded in accordance with the criteria laid down in point 3.6.5[b,c]

[a]The criteria apply independently, i.e. CFS meet one or more of them. [b]Points 3.6.2 to 3.6.5 of Annex II of Regulation (EC) No 1107/2009 [[1]] define hazard-based criteria for substances that must not be approved, so-called cut-off criteria. [c]For endocrine disrupters, currently the interim criteria laid down under point 3.6.5 of Annex II of Regulation (EC) No 1107/2009 [[1]] apply, i.e. substances classified as carcinogenic category 2 and toxic for reproduction category 2.

Faust *et al.*

Faust *et al. Environmental Sciences Europe* 2014 26:11 doi:10.1186/s12302-014 0011-8

As a support for the preparation of the initial list, COM commissioned a contract study to the Food Chain Evaluation Consortium (FCEC). The FCEC delivered their study report in July 2013 [[11]]. Subsequently, the report was presented to the competent authorities of the Member States, and it was made available to all stakeholders via the Commission's CIRCA platform [[12]].Thus, although not formerly published, the report is in the public domain and was accessible for the purposes of this paper via the German Federal Environment Agency (UBA).

It was not the task of the FCEC to set up the initial list of CFS - this is the privilege of COM - but to do the necessary preparatory work, which was

1. to compile the data needed for decision-making from the legally relevant documents, i.e. those documents on which the original decisions for approval of active substances have been based, such as the official Review Reports, EFSA Conclusions, and Draft Assessment Reports, and
2. to explore options for the interpretation and operationalisation of the criteria for the identification of CFS (Table 1), where the legal text and the available data leave room for judgments and where corresponding rules for data assessments had not already been fixed in a corresponding Commission Working Document on 'Evidence Needed to Identify POP, PBT and vPvB Properties of Pesticides' [[13]].

As a consequence, the FCEC report did not directly provide a list of CFS, but it included separate lists of active substances that were considered to fulfil individual CFS criteria or sub-criteria, for example persistence in soil, in sediments, and in water. Where applicable, the report provided various versions of these lists, each representing the outcome of a different interpretation of a legal criterion, such as different measures and trigger values for a 'significantly lower ADI'. In each of these cases the report provided arguments for the option that the authors considered to be the most appropriate one.

Thus, by combining the information from these individual lists, it is possible to obtain a list of potential CFS that have a high chance for becoming actually included in the initial list of CFS that COM is going to establish. This opportunity was used for the purpose of this study.

Comparative Assessments

In the future, EU Member States shall perform a comparative assessment whenever they evaluate any application for authorisation of a PPP that contains a CFS, in the following shortly denoted as candidate product. A comparative assessment may be initiated by an application for the authorisation of a new candidate product, for the renewal of an existing authorisation, or for the amendment of an authorisation for new uses of a candidate product. Comparative assessments must be performed for each use of a candidate product. A candidate product shall not be authorised for a use for which an alternative chemical product or a non-chemical control method is available, if the following requirements are fulfilled (Article 50 in conjunction with Annex IV of the PPP Regulation):

1. Experience from practical use of the alternative is available.
2. The alternative has a comparable efficacy against target pests.
3. The alternative can be used without significant economic or practical disadvantages, including impacts on the so-called minor uses.
4. The substitution does not compromise resistance management and the minimisation of the occurrence of resistance.
5. The alternative product or method is 'significantly safer for human or animal health or the environment'.

Thus, the comparative assessment can be divided into two major parts: a comparative agronomic assessment covering points (i) to (iv) and a comparative safety assessment as required by point (v). This paper focuses on the safety assessment part. In addition, it does not discuss comparisons of chemical PPPs with non-chemical protection methods (e.g. mechanical methods or bio-pesticides such as viruses and bacterial strains).

Annex IV to the PPP Regulation clarifies that the increase in safety that is achieved by a substitution shall be demonstrated in terms of a 'significantly lower risk'. In general, competent authorities shall identify such significant differences in risk 'on a case-by-case basis'. A significance level is not specified for the comparative human health risk assessment but for the environmental risk assessment: 'if relevant, a factor of at least 10 for the toxicity/exposure ratio (TER) of different plant protection products is considered a significant difference in risk' [[1]].

For conducting the agronomic part of the comparative assessments, guidance has been developed by EPPO, the European and Mediterranean Plant Protection Organization [[14]]. For the risk assessment part, COM is currently working on a guidance document, based on a proposal by Sweden [[15]]. This guidance aims to support the Member States, but it will not establish detailed and legally binding rules. Basically, it will be up to the decision of the Member States on how they actually conduct comparative risk assessments.

Regulatory Impact

In order to prepare for the new task of conducting comparative product assessments, competent authorities need to obtain an overview about the dimension of the additional workload they may be confronted with. Earlier assessments of the potential impacts of the new PPP legislation estimated that about 15% to 25% of the approved active substances could become candidates for substitution [[16],[17]], but these studies were based on limited and uncertain databases and they did not further explore the consequences in terms of numbers of potentially affected products and uses of products, and in terms of numbers of pairs or groups of products that could become subject to a comparative risk assessment. However, with the data from the recent FCEC report, a solid basis is now available for such calculations.

Aims and Approach

This study aimed to estimate the potential number of cases for which a comparative assessment will need to be carried out, in terms of (i) the number of CFS-containing candidate products, (ii) the number of different uses of such candidate products, (iii) the number of uses of candidate products for which alternative products are available, and (iv) the possible number of pairwise comparisons of candidate and alternative products for all common uses. For the purpose of this study, we assumed that candidate products are only compared to alternative products that do not contain any CFS. In principle, neither the legal text nor the available draft of the EU Guidance Document excludes the possibility of a substitution of a candidate product by another candidate product, if a significant risk reduction would result. However, we assumed that this would be a rare situation and excluded it from further consideration.

The exercise did not cover the whole European Union but was conducted for the German PPP market as an example. The calculations were based on the simplistic assumption of a static market share of candidate products. This means that authorisations for all uses of all CFS-containing PPPs that are currently on the market would become subject to renewal in the future or that they will be replaced by an equal number of new candidate products with an equal number of uses. In this way we assumed to obtain upper limit estimates of the potential number of cases for comparative assessments that may accumulate over a period of time equal to the average duration of authorisations (typically granted for 10 years).

The identification of CFS results from hazardous properties for humans or for the environment or for both. For human and environmental risk assessments of PPPs, different aims, methods, and procedures apply and they are often performed separately by different institutions. In Germany, for instance, the Federal Institute for Risk Assessment (BfR), the Julius Kühn Institute (JKI), and the Federal Environment Agency (UBA) separately assess risks for humans, for honey bees, and for the environment, respectively. Therefore, we were also interested to see whether the number of cases for comparative risk assessments would be significantly reduced, if comparative environmental assessments would be conducted only for those candidate products that contain CFS as identified by hazardous environmental properties. To this end, we performed all calculations twice: once for all candidate products and once only for those products that contain CFS that have been identified for reasons of environmental hazards (exclusively or in addition to human health hazards). The way by which we achieved this discrimination is explained in the 'Main text' section.

The list of potential CFS resulting from the FCEC report is currently subject to final revisions by the Commission and the Member States. In order to avoid any false discrimination of substances or products in this interim situation, this paper is exclusively focused on the quantitative aspects of the subject, providing numbers of substances, products, and uses, but no names. As a result of the ongoing revisions, the forthcoming official CFS list can be expected to be slightly but not substantially shorter than assumed in this paper [[10]]. Consequently, the figures provided in this paper for the number of candidate products can be expected to represent a slight overestimation of the findings that will be obtainable by performing the same analyses on the basis of the final CFS list.

MAIN TEXT

How Many CFS?

Almost 100 active substances are potential candidates for substitution, which is roughly a quarter of all approved active substances. This derives from an examination of the information provided in the FCEC report to the Commission [[11]], as detailed in the 'Methods' section. The number refers to the list of approved active substances laid down in the Annex to Commission Implementing Regulation (EU) No 540/2011 [[18]], as amended until 31 January 2013. Where the legal definition of an approved active substance includes different varieties of a parent compound or where the approval applies to a defined group of compounds, these were counted as a single entity. Earlier impact studies expected CFS proportions between 15% and 25% [[16],[17]]. Our findings show that these prognoses indicated the dimension of the problem quite accurately.

Potential CFS are spread across all major use categories, as defined in the EU Pesticides database [[19]]. Between 10% and 25% of substances used as fungicides, insecticides, plant growth regulators, or multi-purpose pesticides are potential CFS (Table 2). For acaricides and herbicides, the proportions are higher, with around 40% of the substances being potentially affected. From the small group of rodenticides, even 60% are potential CFS.

Table 2: Proportion of active substances identified as potential CFS, broken down by use categories

Use category[a]	Total number of approved active substances[b]	Proportion of potential CFSc(%)
AC	8	38
FU	103	25
HB	110	41
IN	55	18
PG	25	12
RE	17	0
RO	5	60
Other	14	0
Multi	35	23
Not assigned	6	0
All	378	26

[a]Assignment to use categories as given in the EU Pesticides database [[19]]; AC, acaricides; FU, fungicides; HB, herbicides; IN, insecticides; PG, plant growth regulators; RE, repellents; RO, rodenticides; Other, attractants, bactericides, elicitors, molluscicides, nematicides, and plant activators; Multi, multiple use categories apply to the same substance, also including uses as dessicant in addition to one or more of the other listed categories. [b]Reference date: 31 January 2013; counting of approved active substances refers to the legal definitions listed in Commission Implementing Regulation (EU) 540/2011 [[18]]; from a chemical perspective, parts of these are groups of substances or mixtures of substances; in addition to chemicals, the legal substance definition also includes viruses and bacteria, the so-called bio-pesticides. [c]Percentages rounded to integer values; potential CFS are those identified in the FCEC report [[11]], as detailed in the text.

Faust *et al.*

Faust *et al.* *Environmental Sciences Europe* 2014 26:11, doi:10.1186/s12302-014-0011-8

The so-called PBT properties have an outstanding importance for the number of substances identified as a potential CFS (Table 3). Almost 80% of the potential CFS fulfil the second CFS criterion (see Table 1), which means that they meet two of the criteria to be considered as persistent (P), bioaccumulative (B), and/or toxic (T), as defined under point 3.7.2 of Annex II to the PPP Regulation. Low ADI, ARfD, or AOEL values; reproductive toxicity class 1A or 1B; or endocrine disrupting properties (criteria 1, 6, and 7 in Table 1) have a lower impact on the number of potential CFS, i.e. 22%, 9%, and 7% of the candidates, respectively. Even less influence on the number of CFS has criterion 4 (Table 1), i.e. a significant proportion of non-active isomers: only 2% of the potential CFS meet this criterion. The remaining two criteria 3 and 5 ('nature of critical effects' and carcinogens class 1A/1B) were not found to apply to any approved active substances in the FCEC report. Sixteen percent of all potential CFS were found to meet two or more of the seven identification criteria. There is no obvious association between applicable CFS criteria and the use categories of substances (Table 3). Substances meeting two PBT criteria, for instance, come from all major use categories, whereupon the proportions basically reflect the different sizes of the use groups.

Table 3: Breakdown of the number of potential CFS by use categories and identification criteria

CFS criterion[a]		Use category[e]							Sum
AC		FU	HB	IN	PG	RO	Multi		
1. Low ADI/ARfD/AOEL		0	6	8	2	1	0	5	22
2. Two PBT criteria	T_{AQUA}[b]	3	20	32	8	2	2	3	70
	T_{HUMAN}[c]	0	8	8	1	0	0	3	20
	All	3	23	35	8	2	2	4	77
3. Nature of critical effects		0	0	0	0	0	0	0	0
4. Non-active isomers		0	1	1	0	0	0	0	2
5. Carcinogen 1A/1B		0	0	0	0	0	0	0	0
6. Toxic for reproduction 1A/1B		0	3	5	0	0	1	0	9
7. Endocrine disrupting properties[d]		0	2	4	1	0	0	0	7
All		3	26	45	10	3	3	8	98

[a]The numbering refers to the full legal definition of criteria as given in Table 1; corresponding short descriptions are those used in the EU pesticide database [[19]]; different criteria for identification as a CFS may apply to one and the same substance; therefore figures in lines 'All' do not equal sums of values given in the columns. [b]T_{AQUA}: substances fulfil the toxicity criterion for water organisms laid down under point 3.7.2.3 of Annex II to Regulation (EC) No 1107/2009 [[1]], in addition to the P or the B criterion. [c]T_{HUMAN}: substances fulfil any of the human toxicity criteria defined under point 3.7.2.3 of Annex II to Regulation (EC) No 1107/2009 [[1]] in terms of CMR or STOT RE classifications, in addition to the P or the B criterion. [d]For endocrine disrupters, the interim criteria laid down under point 3.6.5 of Annex II of Regulation (EC) No 1107/2009 [[1]] were applied, i.e. substances classified as carcinogenic category 2 and toxic for reproduction category 2. [e]Assignment to use categories as given in the EU Pesticides database [[19]]; AC, acaricides; FU, fungicides; HB, herbicides; IN, insecticides; PG, plant growth regulators; RO, rodenticides; Multi, multiple use categories apply, also including uses as dessicant, nematicide, or repellent in addition to one or more of the categories AC, FU, HB, IN, PG, and RO.

Faust *et al.*

Faust *et al. Environmental Sciences Europe* 2014 **26**:11, doi:10.1186/s12302-014-0011-8

The seven legal CFS criteria include both human health hazards and environmental hazards, but both aspects are separable, as explained in the following. CFS criteria 1, 5, 6, and 7 (low ADI/ARfD/AOEL, carcinogenicity, reproductive toxicity, and endocrine disrupting properties that may cause adverse effects in humans as defined in the legislation) are exclusively based on tests used for human toxicity assessments. Criterion 3 (nature of critical effects) could be interpreted to include both aspects but was not found to be relevant for the initial list of CFS. Criterion 4 (non-active isomers) clearly has a relevance for both human and environmental hazard and risk assessments. Criterion 2 (two PBT criteria) theoretically comprises three different situations: P + B, P + T, and B + T. Practically, however, the combination P + B has apparently no relevance for the initial CFS list. Hence, criterion 2 always combines an exposure indicator (P or B) with a toxicity indicator (T). As laid down under point 3.7.2.3 of Annex II of the PPP Regulation, two different types of toxicity indicators are applicable: indicators of human toxicity (CMR and STOT RE classifications) and indicators of aquatic toxicity (long-term NOEC), in the following denoted as T_{HUMAN} and T_{AQUA}, respectively. As a consequence, it is possible to distinguish between CFS that meet criterion 2 for reasons of human health protection (T_{HUMAN} in combination with P or B) or for reasons of environmental protection (T_{AQUA} in combination with P or B), or both. In summary of these considerations, CFS identified by environmental hazard criteria are those that meet the T criterion for water organisms in addition to the P or the B criterion, and those that contain a significant proportion of non-active isomers. All other CFS are identified for reasons of human health hazards.

By applying these categorisations, we found that only 19% of the potential CFS are exclusively identified by human health criteria; 81% meet environmental hazard criteria, whereby 27% meet both human and environmental hazard criteria and 54% are exclusively identified for reason of environmental hazards.

How many Candidate Products?

The analysis of the German register of authorised PPPs revealed that 25% of the products contain one or more potential CFS and may hence be considered as potential candidate products (Table 4). This proportion is almost identical to the share of CFS in the number of approved active substances. Interestingly, however, only 67 out of 98 potential CFS were actually found in any PPP on the German market, i.e. around 30% of the CFS currently have no relevance for comparative assessments in Germany. In absolute figures, 351 out of 1,378 authorised PPPs are potential candidate products. Two hundred thirty-seven products contain a potential CFS that has been identified for reason of environmental hazards, which is a share of 17% of all products and 67% of all potential candidate products. Thus, on the level of products on the German market, the importance of environmental hazard criteria for the number of candidates is only slightly lower than seen on the level of active substances, where the corresponding fraction is 81% of all potential CFS.

Table 4: Proportion of candidate products containing potential CFS authorised in Germany

Use category[a]	Total number of authorised PPPs[b]	Proportion of candidate PPPs containing one or more potential CFS[c]	
		Containing any potential CFS (%)	Containing CFS identified by environmental hazard criteria[d](%)
Herbicides	567	25	21
Fungicides	308	45	32
Insecticides	281	25	6
Acaricides	109	12	2
Plant growth regulators	59	14	7
Molluscicides	58	0	0
Rodenticides	41	2	2
Glue, sealing wax	31	0	0
Repellents	29	0	0
Sprout inhibitors	18	0	0
Bactericides	9	11	11
Pheromones	3	0	0
Viricides	1	0	0
Nematicides	1	0	0
All[b]	1,378	25	17

As of May 2013. [a]Categorisation as given in the BVL database on plant protection products [[20]]; this categorisation of products is largely but not entirely consistent with the categorisation of active substances used in the EU pesticide database [[19]] (see preceding tables). [b]The given categorisation of products is non-exclusive, i.e. multiple use categories may apply to a single product; this is the case for around 10% of all PPP; as a consequence, the overall number of 1,378 PPPs is smaller than the sum of values in the column. [c]Percentages rounded to integer values; potential CFS are those identified in the FCEC report [[11]], as detailed in the text. [d]Active substances that fulfil the toxicity criterion for water organisms laid down under point 3.7.2.3 of Annex II to Regulation (EC) No 1107/2009 in addition to the P or the B criterion as defined under points 3.7.2.1 and 3.7.2.1 of the same Annex II [[1]], and active substances that contain a significant proportion of non-active isomers; human health criteria may apply additionally.

Faust *et al.*

Faust *et al. Environmental Sciences Europe* 2014 **26**:11, doi:10.1186/s12302-014-0011-8

The potential candidate PPPs are spread across seven major use categories, with shares ranging between 2% and 45%. This is essentially the same situation as observed on the level of CFS. Interestingly, however, the proportions can be quite different for the same use category. In the small group of rodenticides, for instance, 60% of the active substances were found to be potential CFS, but no more than 2% of the rodenticidal products authorised in Germany actually contain any of these CFS. The opposite situation does also occur: In the large group of fungicides, for instance, 25% of the active substances are potential CFS, but 45% of the products on the German market contain one or more of these CFS. The high proportion of 45% potential candidate products amongst fungicides is exceptional; for all other use categories, the fractions are at or below the average of 25%.

How Many Uses of Candidate Products?

On average, potential candidate products are authorised for a broader spectrum of uses than other PPPs. As a consequence, they may necessitate comparative assessments for a disproportionally high number of uses; they account for no more than 25% of all products (see above), but they are authorised for around 50% of all uses. This is the essential outcome of an analysis detailed in the following.

Authorised uses of PPPs are defined in terms of a combination of a crop and a pest from which the crop shall be protected. A small number of other treatment objects (such as food storage rooms) and treatment aims (such as plant growth regulation) are subsumed here also under the terms crop and pest, respectively. Using the status of May 2013, a total of 1,378 PPPs was authorised in Germany for a total of 3,606 uses, defined as different combinations from of a total of 309 crops and 477 pests (Table 5). The fraction of 351 potential candidate products (25% of all products) was authorised for a total of 1,863 uses (52% of all uses), defined as different combinations from 209 crops (68% of all crops) and 264 pests (55% of all pests). The sub-fraction of 237 potential candidate products that contain CFS identified by environmental hazard criteria (17% of all products) was authorised for a total of 1,501 uses (42% of all uses), defined as different combinations from 186 crops (60% of all crops) and 228 pests (48% of all pests).

Table 5: Number of plant protection products, protected crops, controlled pests, and authorised uses in Germany

Parameter			**All PPPs**	**Candidate PPPs containing one or more potential CFS**	
Containing any potential CFS			**Containing CFS identified by environmental hazard criteria**[a]		
Products	Total number		1,378	351	237
Crops[b]	Total number of different crops for which use of PPPs has been authorised		309	209	186
	Number of different crops for which use of an individual PPP has been authorised	Min	1	1	1
		Max	85	85	85
		Median	4	4	4
		Mean[e]	7	7	7
Pests[c]	Total number of different pests against which use of PPPs has been authorised		477	264	228
	Number of different pests against which use of an individual PPP has been authorised	Min	1	1	1
		Max	64	39	39
		Median	3	4	4
		Mean[e]	5	5	5
Uses[d]	Total number of different uses for which PPPs have been authorised		3,606	1,863	1,501
	Number of different uses for which an individual PPP has been authorised	Min	1	1	1
		Max	337	337	337
		Median	6	8	9
		Mean[e]	13	18	18

As of May 2013. Broken down by potential CFS content of products. [a]As defined in the corresponding footnote to Table 4. [b]Including other authorised treatment objects such as food storage rooms for instance. [c]Including other authorised treatment aims such plant growth regulation for instance. [d]Defined by a combination of a crop (or another treatment object) and a pest from which the crop shall be protected (or another treatment aim). [e]Arithmetic mean rounded to integer values.

Faust *et al.*

Faust *et al. Environmental Sciences Europe* 2014 26:11, doi:10.1186/s12302-014-0011-8

On average, every PPP is authorised for use against five different pests in seven different crops. For potential candidate products, the same mean values apply. On the level of uses, however, the situation is different: while the mean number of authorised uses is 13 for all PPPs, potential candidate products have an average of 18 different authorised uses.

How Many Alternatives are Available?

For 767 different uses of potential candidate products, at least one CFS-free alternative product is authorised (Table 6). This is slightly more than 40% of all 1,863 uses of potential candidate products. For around 9%, even more than 10 different alternative products are authorised. For the sub-fraction of potential candidate products that contain CFS identified by environmental hazard criteria, the corresponding proportions are slightly higher: at least one alternative is available for 687 of 1,501 uses, i.e. almost 46%. For almost 12%, even more than 10 different alternative products are authorised.

Table 6: Potential availability of alternative products for authorised uses of CFS-containing candidate products in Germany

Parameter	**All potential candidate products (n= 351)**	**Potential candidate products containing CFS identified by environmental hazard criteria[a] (*n*= 237)**
Availability of alternative products for uses of candidate products	Number of uses	
No alternatives available	1,096	813
Any alternatives available	767	687
1 alternative available	220	162
2 alternatives available	126	112
3 to 5 alternatives available	172	163
6 to 10 alternatives available	82	77

11 to 20 alternatives available	112	118
21 to 50 alternatives available	42	44
51 to 100 alternatives available	12	10
More than 100 alternatives available	1	1
Average spectrum of uses of candidate products	Mean number of uses[b]	
Including all uses of candidate products	18	18
Including only uses of candidate products for which alternatives are available	7	8
Average availability of alternatives	Mean number of alternative products[b]	
For all uses of candidate products	3	4
For all uses of candidate products for which alternatives are available	7	8

Scenario based on the status of authorised products in May 2013. Candidate products are products containing one or more potential CFS; alternatives are products containing no potential CFS. [a]As defined in the corresponding footnote to Table 4. [b]Arithmetic mean rounded to integer values.

Faust *et al.*

Faust *et al. Environmental Sciences Europe* 2014 **26**:11, doi:10.1186/s12302-014-0011-8

On average, one or more alternative products are authorised for around 7 out of 18 different uses of a potential candidate product. For each of these seven uses, the mean number of available alternative products is also around seven. For the sub-fraction of potential candidate products that contain CFS identified by environmental hazard criteria, these mean values differ only marginally.

How many cases for comparative assessments?

In principle, every application for authorisation of a candidate product constitutes a case for comparative assessments, as required by the PPP Regulation. While the assessments may be stopped in an early phase, if alternatives are not available for any of the uses of the candidate product, our analysis revealed that only for 6 out of 351 potential candidate products no alternative products are authorised for any of their uses. The remaining 345 potential candidate products (around 98%) may trigger more detailed comparative assessments, at least of the agronomic aspects, and where appropriate also of the safety aspects. For the sub-fraction of potential candidate products that contain CFS identified by environmental hazard criteria, the proportion is basically the same (Table 7).

Table 7: Potential number of cases for comparative risk assessment of plant protection products in Germany

Cases defined in terms of...		**Counts**	
For products containing any potential CFS		**For products containing CFS identified by environmental hazard criteria[a]**	
Products[b]	Number of all candidate products	351	237
	Number of candidate products for which alternatives are available for one or more of their uses	345	232
Uses	Number of all authorised uses of candidate products	1,863	1,501
	Number of authorised uses of candidate products for which alternative products are available	767	687
Products × Uses	Number of all candidate products times average number of uses[c]	6,232	4,175
	Number of all candidate products times average number of uses for which alternatives are potentially available[c]	2,569	1,910

Products × Uses × Alternatives	Number of all possible pairwise risk comparisons of candidate products with alternative products for all common uses[d]	18,479	15,287

Scenario based on the status of authorised products in May 2013. [a]As defined in the corresponding footnote to Table 4. [b]Candidate products are products containing one or more potential CFS; alternatives are products containing no potential CFS. [c]Calculated with non-rounded values for mean numbers of uses. [d]Calculated by multiplying the number of all candidate products with the average number of uses for which alternatives are potentially available and the mean number of such alternatives for every use; non-rounded mean values were used for the calculations.

Faust *et al.*

Faust *et al. Environmental Sciences Europe* 2014 26:11, doi:10.1186/s12302-014-0011-8

In terms of affected uses, more or less detailed comparative assessments may be required for the 767 different uses of potential candidate products for which at least one alternative product is currently authorised. However, even for the same use, not all candidate products will become subject to authorisation or re-authorisation at the same time. Whenever an authorisation is requested for a single candidate product, comparative assessments will have to be conducted for all its uses. This means that comparative assessments for the same use may have to be conducted repeatedly over time. Therefore, as an additional indicator of the potential workload for competent authorities, the number of potential candidate products can be multiplied with the average number of uses for which alternatives are potentially available. This yields an indicative figure of roughly 2,500 cases of use-specific assessments of candidate products that may become necessary during the next years. For the sub-fraction of potential candidate products that contain CFS identified by environmental hazard criteria, the figure is roughly 25% smaller.

It must be noted that our analysis of uses was based on the exact wording of the use descriptions in the German PPP register. Hence, we did not explore the potential for substituting a candidate product that is only used for a specific pest and/or crop by an alternative product that is authorised for broad-spectrum use against a large group of pests and/or in multiple crops. This could further increase the number of comparative product assessments.

In addition to the number of uses, the number of available alternative products must be taken into account. The broader the spectrum, the higher will be the workload for comparatively assessing all products that are available for a given use. As a further workload indicator, we therefore derived the product of all three factors: The number of potential candidate products × the average number of uses for which alternatives are potentially available × the average number of alternative products that is available for each of these uses. This results in roughly 18,500 cases. This figure is the number of all possible pairwise comparisons of candidate products with alternative products for all common uses of two products. For the sub-fraction of potential candidate products that contain CFS identified by environmental hazard criteria, the figure is roughly 20% smaller.

DISCUSSION

Estimated case numbers in this paper are based on a draft CFS list and on a simplistic static scenario, reflecting neither fluctuations in the number of authorised products nor any potential future trends. Therefore, the figures should not be taken as precise estimates but rather as indications of the expectable dimension of the upcoming demands on comparative assessments. In any case, the results of the present study indicate that several thousands of cases might need to be evaluated. Hence, the new task of performing comparative assessments of PPPs may pose a formidable challenge to competent authorities. Not only are the quantitative demands considerable, the established assessment methodologies for PPPs and the existing organisational arrangements are also only designed for checking compliance of individual PPPs with legal acceptability criteria. They were not developed for comparing products with each other. New data handling systems, assessment procedures, and decision rules might hence be required for efficiently dealing with the new task.

The comparative agronomic assessment and the comparative safety assessment may be performed either subsequently or in parallel. The EPPO guidance on comparative assessment proposed that the agronomic assessment should be carried out first [[14]], and the latest draft of the forthcoming Commission guidance document suggests the same [[15]]. With the aim of optimal efficiency of the overall procedure, this appears to be a rational and self-evident approach, as the agronomic assessment can be expected to provide a strong filter. Thus, the novel and demanding task of comparing the human and environmental risks of different products would be limited to the lowest possible number of cases. On the other hand, however, it must be taken into consideration that the agronomic part of the assessment includes criteria that need further interpretation and specification, in particular the requirement for no 'significant' economic or practical disadvantages. Corresponding definitions of significance may be based on isolated agronomic considerations, but it is also imaginable, that the assessment of the significance of economic disadvantages should take account of potential advantages for human health and the environment. Thus, the number of cases for comparative human and environmental risk assessments will depend on the detailed criteria and procedural arrangements of the overall process.

The workload for performing a complete set of comparative human and environmental risk assessments for all the uses of a candidate product will certainly be quite different, depending on the respective product. These different assessment situations result from the different numbers of uses of potential candidate products, which vary between one and a few hundreds, and also from the different numbers of potentially available alternatives, which range between zero and more than a hundred (Tables 5 and 6). In addition, the initial comparative assessments of substances available for a certain common use may be more demanding than a repeated assessment that is triggered by an application for authorisation of another product for the same use. Furthermore, the risks resulting from different uses of the same product may be quite similar or even identical, and most likely there will also be situations where the available spectrum of candidate products and alternative products is identical for different uses. Thus, a more detailed assessment of the workload from comparative risks assessments and the development of an optimal strategy for efficiently dealing with the issue would need to consider these different types of assessment situations. Such analyses will become possible as soon as the final official CFS list becomes available.

Besides the need to further clarify these details, the high number of expectable cases calls for the development of an electronic decision support instrument which would enable competent authorities to rapidly carry out initial comparative risk assessments. Such a tool should allow to filter out clear-cut cases in a semi-automatic manner and to separate them from borderline cases which require in-depth expert judgment. To this end, the legal requirement for a significant risk reduction needs to be translated into well-defined programmable decision rules. In addition, an efficient application of such a computer-based decision support tool depends on the availability of a continuously updated database with all regulatory relevant information on the risks of individual products that is required for the comparative assessments. In Germany, regulatory risk assessment reports for individual PPPs currently exist in the form of text files only, and the same may presumably apply to most other EU Member States. For performing a comparative assessment, all the relevant risk indicators, such as TER values for the various ecotoxicological endpoints, must therefore be compiled manually from the individual assessment reports. Improving this situation is essential for a less time-consuming future practice. In addition, efforts should be made to identify representative endpoints or to define suitable indicators for facilitating decision-making within the process, e.g. for sorting products according to their risk. Within a broader scope, cooperation between Member States authorities may be considered as a further means for efficiently dealing with the large numbers of comparative risk assessments needed. Comparative risk assessment of PPPs is clearly a national task, but a zonal authorisation of PPPs across different Member States has been introduced with the new PPP Regulation. In this situation, sharing the workload for comparative risk assessments between Member States in the same zone seems to be a self-suggesting option that deserves further exploration.

CONCLUSIONS

The expectable number of cases for comparative assessments of PPPs is high. In Germany, it may comprise up to a quarter of all products and half of all uses of products. This puts regulatory authorities under considerable pressure to develop appropriate strategies for efficient handling of the task.

METHODS

Data presented in this study were generated by systematically compiling and interlinking information from three different sources:

The FCEC report to the European Commission [[11]] for information on CFS properties of approved active substances (approval status as of 31 January 2013)

The EU Pesticides database [[19]] for allocating approved active substances to use categories

The database on Authorised Plant Protection Products of the German Federal Office of Consumer Protection and Food Safety (BVL) [[20]] for information on the number, nature, active ingredients, uses, and use categories of authorised PPP in Germany (authorisation status as of May 2013)

A list of potential CFS was derived from the FCEC report by aggregating relevant information as follows:

CFS criterion 1 (low ADI/ARfD/AOEL): Separate assessments of ADI, ARfD, and AOEL values (Tables A1, A2, and A3 of the FCEC report) were merged under the assumption that the following decision rule applies: values are considered to fulfil the criterion when they are below the 5% percentile of a use group as defined in the EU pesticides database.

CFS criterion 2 (two PBT criteria): Separate assessments of half-life in water, sediments, and soil (Tables A5, A7, and A9 of the FCEC report) were merged for assessments of persistence (P). Assessments of bioaccumulation (B) were directly retrieved (from Table A10 of the FCEC report). Assessments of aquatic toxicity (T_{AQUA}) were generated by merging separate assessments of the toxicity to fish, algae, daphnids, and other aquatic species (as provided in Tables A11, A12, A13, and A14 of the FCEC report). Assessments of human toxicity (T_{HUMAN}) were obtained by merging the separate information on CMR and STOT RE classifications (provided in Tables A16, A20, A22, A24, A26, and A 27 of the FCEC report).

CFS criterion 4 (non-active isomers): Assessments were directly abstracted (from Table A15 of the FCEC report).

CFS criterion 6 toxic for (reproduction 1A/1B): Separate information on existing and forthcoming classifications of reproductive toxicity was merged (from Tables A22 and A23 of the FCEC report).

CFS criterion 7 (endocrine disrupting properties): For assessments of endocrine disrupting properties according to the interim criteria laid down in Regulation (EC) No 1272/2008, i.e. substances that are classified as both carcinogenic category 2 and toxic for reproduction category 2, separate information on these two properties (in Tables A21 and A25 of the FCEC report) were combined accordingly.

CFS criteria 3 and 5 ('nature of critical effects' and carcinogen 1A/1B): No substances fulfilling any of these criteria were identified in the FCEC report.

AUTHORS' CONTRIBUTIONS

All authors made substantial contributions to the conception and design of the study and the acquisition of data. SR and CV performed the database analyses. MF drafted the manuscript. All authors commented and agreed on the final version.

ACKNOWLEDGEMENTS

Funding by the German Federal Environment Agency (FKZ 3712 67 406) is gratefully acknowledged.

Disclaimer

The findings and conclusions in this paper are those of the authors and do not necessarily represent the view of the German Federal Environment Agency (UBA).

1. (2009) Regulation (EC) No 1107/2009 of the European Parliament and of the Council of 21 October 2009 concerning the placing of plant protection products on the market and repealing Council Directives 79/117/EEC and 91/414/EEC. Off J Eur Union L 309:1-50

2. (1991) Council Directive of 15 July 1991 concerning the placing of plant protection products on the market (91/414/EEC). Off J Eur Communities L 230:1-32
3. (2012) Regulation (EU) No 528/2012 of the European Parliament and of the Council of 22 May 2012 concerning the making available on the market and use of biocidal products. Off J Eur Union L 167:1-123
4. (2009) Directive 2009/128/EC of the European Parliament and of the Council of 21 October 2009 establishing a framework for Community action to achieve the sustainable use of pesticides. Off J Eur Union L 309:71-86
5. [www.kemi.se] KEMI (Swedish Chemicals Agency): The substitution principle. KEMI Report Nr. 8/07.2007. Available online at the agencies website: .
6. [https:/ / www.vci.de/ Downloads/ Publikation/ 28_Reform-europaeischen-Chemikalien rechts.pdf] VCI (Verband der Chemischen Industrie) (Association of chemical industries): Reform des europäischen Chemikalienrechts - Fakten, Meinungen, Hintergründe. 2005. (Reform of the European chemicals legislation – facts, opinions, background; in German). 2005. Available online at .
7. Commission of the European Communities: White Paper strategy for a future chemicals policy. 2001. Brussels, 27.2.2001, COM(2001) 88 final.
8. (2006) Regulation (EC) No 1907/2006 of the European Parliament and of the Council of 18 December 2006 concerning the Registration, Evaluation, Authorisation and Restriction of Chemicals (REACH), establishing a European Chemicals Agency, amending Directive 1999/45/EC and repealing Council Regulation (EEC) No 793/93 and Commission Regulation (EC) No 1488/94 as well as Council Directive 76/769/EEC and Commission Directives 91/155/EEC, 93/67/EEC, 93/105/EC and 2000/21/EC. Off J Eur Union L 396:1-849
9. [http:/ / www2.unitar.org/ cwm/ publications/ cw/ np/ np_pdf/ Sweden_National_Profile.pdf] KEMI (Swedish Chemicals Agency): Management of chemicals and pesticides in Sweden - national profile. 2000. Available online at: .

10. [http:/ / www.pesticides.gov.uk/ guidance/ industries/ pesticides/ topics/ pesticide-approvals/ eu/ european-regulation/ Regulation+_EC_No_1107_2009-progres s_on_endocrine_ disrupters_and_candi dates_for+substitution] HSE (Health and Safety Executive): Regulation (EC) No 1107/2009: progress on endocrine disrupters and candidates for substitution, Information Update: 01/2014. 2014. Issued: 17th January 2014 online at .
11. FCEC (Food Chain Evaluation Consortium): Ad-hoc study to support the initial establishment of the list of candidates for substitution as required in Article 80(7) of Regulation (EC) No 1107/2009. 2013. Framework Contract for evaluation and evaluation related services - Lot 3: Food Chain, Final report submitted to the European Commission, Directorate General for Health and Consumers, European Commission DG SANCO, Rue de la Loi 200, 1049 Brussels, 09.07.2013. Available online to Member States authorities and registered stakeholders via the Commission's CIRCA platform.
12. [https://circabc.europa.eu] European Union: CIRCA (Communication & Information Resource Centre for Administrations, Businesses and Citizens). 2014. .
13. [http:/ / ec.europa.eu/ food/ plant/ pesticides/ approval_active_ substances/ docs/ wd_evidence_needed_to_identify_pop_ pbt_ vpvb_properties_rev3_en.pdf] COM (European Commission, Health and Consumers Directorate General): DG SANCO working document on "Evidence needed to identify POP, PBT and vPvB properties of pesticides". 2012. Brussels, 25.09.2012 – rev. 3. Available online at .
14. (2011) Guidance on comparative assessment. Bull OEPP/EPPO Bull 41:256-259
15. COM (European Commission, Health and Consumers Directorate General): Draft Guidance document on comparative assessment and substitution of plant protection products in accordance with Regulation (EC) No 1107/2009. 2014. SANCO/11507/2013 rev. 8, March 2014. Available to stakeholders and competent Member States authorities.
16. [http://www.nefyto.nl/6008/Impact-report-final-PSD-UK-%28May-2008%29.pdf] PSD (Pesticide Safety Directorate, United Kingdom): Assessment of the impact on crop protection

in the UK of the 'cut-off criteria' and substitution provisions in the proposed Regulation of the European Parliament and of the Council concerning the placing of plant protection products in the market. 2008. Online available at .

17. [http://www.dinamica-fp.it/centri/ra/docs/allegatoENEA.pdf] Rapagani MR, Maglulio M, Piccolo M, Nencini L, Galassi T, Mazzini F: Future availability of pesticides in the integrated pest management agricultural programme in Italy in accordance with the application of the new European Regulation No. 1107/2009 concerning the placing of plant protection products on the market: impact of the application of cut-off criteria and selection criteria for substances that are candidate for substitution. 2011. ENEA - Agenzia Nazionale per le Nuove Tecnologie, L'Energia e lo Sviluppo Economico Sostenibile Future, Report RT/2011/8/ ENEA, Roma, Italy, 2011. Available at .
18. (2011) Commission Implementing Regulation (EU) No 540/2011 of 25 May 2011 implementing Regulation (EC) No 1107/2009 of the European Parliament and of the Council as regards the list of approved active substances. Basic Legal Act published in Off J Eur Union L L 153:1-186Regulation as amended until 31 January 2013
19. [http://ec.europa.eu/sanco_pesticides/public/] European Union: EU Pesticides database. 2014. .
20. [http://www.bvl.bund.de] BVL (Bundesamt fürVerbraucherschutz und Lebensmittelsicherheit) (German Federal Office of Consumer Protection and Food Safety): Database on authorised plant protection products. 2014. Available upon subscription as detailed on the agencies website at .

Chapter 9

The use of Adjuvants to Improve uptake of Phosphorous Acid Applied to Pinus Radiata needles for Control of Foliar Phytophthora Diseases

Carol Rolando[1], Robyn Gaskin[2], David Horgan[2], Nari Williams[1], and Martin K-F Bader[1]

[1]Scion, 49 Sala Street, Fenton Park, Private Bag 3020, Rotorua 3046, New Zealand

[2]Plant Protection Chemistry NZ, 49 Sala Street, Fenton Park, Rotorua 3043, New Zealand

ABSTRACT

Background

Phosphorous acid is being investigated as a fungicide for the management of a needle disease caused by *Phytophthora pluvialis* in *Pinus radiata* in New Zealand. However, little is known about the penetration characteristics of this fungicide into *Pinus radiata* foliage. This study was undertaken to determine: i) the penetration characteristics of a commercial phosphorous acid formulation, applied at 3 kg ha^{-1} and 12 kg ha^{-1} in 100 L water, into *Pinus radiata* foliage and, ii) the effect of four commercially available adjuvants on phosphorous acid uptake into *Pinus radiata*foliage. Efficacy of the best treatment was tested in vitro with two *Phytophthora* species,*Phytophthora kernoviae* and *P. pluvialis*.

Methods

Foliar uptake of orthophosphoric acid radio-labelled with [^{32}P] was used to determine penetration characteristics of phosphorous acid as affected by the adjuvants tested. Needles collected from potted *Pinus radiata* plants sprayed with the best performing treatment were used in a detached needle assay, where treated needles were exposed to the zoospores of the *Phytophthora* species being tested.

Results

Uptake of the phosphorous acid formulation into *Pinus radiata* needles applied without adjuvants was low (3.9-6.6%) at both concentrations tested. An alcohol ethoxylate adjuvant (applied at 0.2%) and two organosilicone adjuvants (applied at 0.2%) were found to significantly increase uptake of the fungicide over that applied alone or in combination with pinolene (applied at 0.35%). Highest uptake of phosphorous acid into *Pinus radiata* foliage (51.6%) occurred over a period of 72 hours when the phosphorous acid was applied at the equivalent of 12 kg ha^{-1} with an organosilicone blend adjuvant at 0.2%. A significant reduction in lesion length on infected needles was found relative to the control when the two phosphorous acid treatments were applied with 0.2% organosilicone blend adjuvant. However, there were no significant differences in lesion length between the two treatments.

Conclusions

These results show that an organosilicone adjuvant significantly enhances uptake of phosphorous acid into *Pinus radiata* needles and may increase its efficacy against *Phytophthora* species.

BACKGROUND

A new needle disease was reported in *Pinus radiata* D. Don stands on the east coast of New Zealand during routine forest surveillance in late April 2008. Observations of the symptoms, their development and the apparent spread of the disease, known locally as red needle cast, all gave a strong indication of a likely biotic cause, which led to further investigations. Isolations made from symptomatic needles yielded an undescribed species of *Phytophthora* as the likely causal pathogen (Dick et al. [2014]). The causal pathogen has recently been described as *Phytophthora pluvialis*which is considered to be endemic to Oregon, USA (Reeser et al. [2013]). Subsequently, a second species *Phytophthora kernoviae* Brasier, Beales & Kirk, was also recovered from needles, though at a consistently lower rate (Dick et al. [2014]). Concurrent to epidemiological work for red needle cast, a range of trials were initiated to investigate possible fungicides that could be effective for management of red needle cast on *Pinus radiata* in New Zealand. Phosphorous acid was chosen as a key fungicide for further research on potential red needle cast management since this active ingredient (in the form of phosphorous acid) is used widely in agriculture and horticulture for control of *Phytophthora* species (Lobato et al. [2011]; Rebollar-Alviter et al. [2007]; Shearer and Crane [2009]).

Phosphorous acid fungicides, applied either pre-infection or for protectant activity, are the only known ambi-mobile fungicides capable of moving upward and downward in plants (Cohen and Coffey [1986]). Phosphorous acid exhibits a complex mode of action, acting both directly on the pathogen and also indirectly by stimulating host defence responses to ultimately inhibit pathogen growth (Tomlin

[2006]). The fungicide can persist in host plants for several months post application, however persistence is host dependent (Tynan et al. [2001]). When treated, most plants exhibit phytotoxic effects above a (species dependent) threshold concentration, with field treated plants more sensitive to phytotoxic effects than those treated in a glasshouse (Hardy et al.[2001]). Kanaskie et al. ([2011]) showed the potential for phosphorous acid to control outbreaks of sudden oak death, caused by *Phytophthora ramorum* S. Werres, A.W.A.M. de Cock & W.A. Man, in*Notholithocarpus densiflorus* (Hook. & Arn.) Manos, Cannon & S.H. Oh (south-western Oregon tanoak). Besides the work described by Kanaskie et al. ([2011]), there are few reports describing the use of phosphorous acid to control *Phytophthora* spp. associated foliar diseases in mature trees. This is likely because there are few tree-infecting aerial *Phytophthora* spp. currently known and those which are known (such as *P. ramorum, P. kernoviae, P. pinifolia* and *P. pluvialis*) have been relatively recently observed. Graham ([2011]) suggested options for using phosphorous acid to control *Phytophthora* species in forest trees based on work done on citrus orchards. Phosphorous acid has been successfully used to control *Phytophthora* root rot in *Pinus radiata*nurseries (Reglinski et al. [2009]). However, the results of this study cannot readily be extrapolated to airborne *Phytophthora* pathogens infecting mature trees.

Maximising uptake of pesticides is important to optimise efficacy and cost-effectiveness of any chemical control strategy. However, little is known about the penetration characteristics of phosphorous acid into *Pinus radiata* foliage. Agrochemical adjuvants are commonly used to improve efficacy of systemic pesticides by improving spray deposition and retention on the target species and/or by increasing the penetration of the pesticides into the plant (Zabkiewicz [2000]).

This study was undertaken to determine: i) the penetration characteristics of a commercial phosphorous acid formulation into *Pinus radiata* foliage and, ii) the effect of commercially available adjuvants on phosphorous acid uptake into *Pinus radiata* foliage. Efficacy of the best treatment was subsequently tested in an *in-vitro* bioassay conducted with the pathogens, *Phytophthora kernoviae* and *P. pluvialis*. Selection was based on a combination of adjuvant chemistry, their mode of action, and current relative cost. Four adjuvants, representing three classes of adjuvant, were selected for testing: i) a low-cost pinolene that is used as a wetter, sticker and rainproofer, ii) a moderately priced non-ionic surfactant typically used to aid wetting, sticking and penetration of fungicides, and iii) two relatively expensive organosilicones that are known to improve retention and coverage of sprays, and are often utilised to increase uptake of systemic pesticides into plants.

METHODS

Three clones (referred to as A, B and C) of grafted *Pinus radiata,* hereafter referred to as the potted plants, known to be susceptible to the pathogens *Phytophthora kernoviae* and *P. pluvialis,*were used in this investigation. The potted plants, supplied from the Radiata Pine Breeding Company Ltd. (Rotorua, New Zealand), were obtained in mid-autumn from the Scion nursery (Rotorua, New Zealand) and were equilibrated for two weeks in a controlled environment cabinet (20/15°C, 70% relative humidity, 12 h photoperiod, 500 µmol m^{-2} sec^{-1} light) prior to treatment application. All treatments were applied to one-year-old needles located on the main leader of the potted plants. There were three potted plants of each clone for each treatment, such that all treatments were replicated three times, with three experimental units per clone (three clones x three potted plants of each clone).

A preliminary experiment was undertaken to gauge the uptake of phosphorous acid into *Pinus radiata* needles without the use of an adjuvant. As there is no label rate for aerial application of phosphorous acid to mature *Pinus radiata* plantations in New Zealand, the rates used were obtained from aerial application rates used in Western Australia, for *Lambertia inermis* var. *inermis*and *Banksia grandis,* and North America for *N. densiflorus* (Kanaskie et al. [2011]; Shearer and Crane [2009]). Phosphorous acid (Agrifos ® 600, Key Industries; 600 g L^{-1} phosphorous acid) was applied in 100 L ha^{-1} of water at two rates (3 kg ha^{-1} and 12 kg ha^{-1}), Treatments 1 & 2, Table 1. Based on the results of this initial test, it was decided to use the highest application rate of phosphorous acid in further experiments to increase the likelihood of active ingredient being absorbed by the foliage.

Table 1: Uptake of [^{32}P] phosphorous acid into*Pinus radiata*needles with or without various adjuvants at 24 hours after treatment (HAT) and droplet spread on needle surface

Treatment no	Rate of active (kg ha^{-1})	Adjuvant	Adjuvant concen tration	[^{32}P] uptake (%)	[32P] uptake dose* (kg ha^{-1})	Spread area of droplets (mm^2)
1	3	None	-	6.6 cd	0.20	0.8 h
2	12	None	-	3.9 d	0.47	1.0 g
3	12	Pinolene	0.35	6.8 cd	0.82	1.2 f
4	12	Alcohol ethoxylate	0.05	7.7 cd	0.92	1.5 e
5	12	Alcohol ethoxylate	0.10	18.6 b	2.23	1.7 d
6	12	Alcohol ethoxylate	0.20	14.2 bc	1.70	2.0 c
7	12	Organosilicone penetrant	0.20	20.1 b	2.41	2.3 b
8	12	Organosilicone blend	0.20	36.9 a	4.43	2.7 a

Means within columns sharing common postscripts are not significantly different.

*Nominal dose calculated as (phosphorous acid rate x %_uptake).

Rolando *et al.*

Rolando *et al. New Zealand Journal of Forestry Science* 2014 44:8 doi:10.1186/s40490-014-0008-5

Experiment 1: Effect of Four Adjuvants on Uptake of Phosphorous Acid

Description of Adjuvants

The adjuvants tested were i) a pinolene (Nu-Film-17 ®, Key Industries Ltd, Auckland), ii) a linear alcohol ethoxylate (950 g L^{-1}, Actiwett ®, Etec Crop Solutions Ltd, Auckland), iii) a pH-stable organosilicone polymer with super-penetrant activity (Gaskin et al. [2012]) containing alcohol ethoxylate and polyalkylene compounds (Du-Wett ® Stainless,

Etec Crop Solutions Ltd, Auckland) and iv) an organosilicone-blend containing siloxane polyalkeneoxide copolymers (Du-Wett ®, Etec Crop Solutions Ltd, Auckland) designed for use as an horticultural super-spreader (AgriMedia Ltd[2013]). All adjuvants used in the trial were tested within manufacturer's recommended rates or at rates known to be effective for penetration into horticultural and agricultural crops.

Determination of Treatment Penetration

The penetration characteristics of phosphorous acid were tested for up to 24 h after treatment (HAT). Phosphorous acid was applied as Agrifos ® 600, at the equivalent of 12 kg ha^{-1} in 100 L ha^{-1} water in combination with the four commercially available adjuvants described in 2.1.1. Details of the adjuvant concentrations (Treatments 3–8) are given in Table 1.Orthophosphoric acid radio-labelled with [^{32}P] (ARC, Inc., USA) was added to freshly prepared treatment solutions in deionised water 30 minutes prior to use to enable the radiaoactive chemical to equilibrate with the mono-and di-potassium salts present in the commercial phosphorous acid formulation. The [^{32}P] phosphoric acid comprised <1% mass of the total phosphorous acid in solution at all times. Droplets (10 × 0.24 µL) of the radiolabelled treatment solution were applied using a microsyringe to the central region of a single needle per fascicle such that they simulated a 100 L ha^{-1} spray application volume (i.e. 1 µL cm^{-2}). Treated fascicles were pre-tagged with coloured electrical wire for identification.

The quantity of radiolabelled phosphorous acid applied to each needle in each treatment was determined by dispensing droplets, as applied to plants, directly into scintillation vials (3 replicates). Following treatment application, needle fascicles were harvested at 24 HAT (Table 1). The entire fascicle, including the treated leaf surface, was washed with a mixture of water and ethanol (1:1 v/v; 2 × 4 mL) to recover any unabsorbed phosphorous acid. Prior audits determined that such wash-offs could recover >95% of applied phosphorous acid solutions from foliage surfaces at dry-down. Foliage in Treatment #3 (Table 1) was given an additional wash with 50% aqueous acetone to ensure all phosphorous acid remaining on the surface was recovered. This extra wash was carried out because of the low solubility of the pinolene adjuvant in aqueous solution.

Scintillant solution (13 mL ACS; Amersham Biosciences, Australia) was added to the washings and radioactivity was quantified by liquid scintillation counting (Packard Tricarb 2100 TR). Foliar uptake was defined as the radioactivity not recovered from washing the treated needles and was calculated as a percentage of the applied dose.

Determination of Droplet Spread

In a separate test, all treatment solutions (1–8; Table 1) were prepared as for the penetration experiment but with the addition of a water-soluble fluorescent dye, (Blankophore P, Bayer) at a concentration of 0.5% w/v immediately prior to droplet application. Twelve droplets (0.24 μL) of each solution were applied to each of three needles on each potted plant (and different clones). Droplets were left to spread for 10 minutes. Needles were then illuminated with ultra-violet light and photographed. The area of droplet spread in each case was determined using image analysis software (V++™, Digital Optics Ltd).

Experiment 2: Optimising Uptake of Phosphorous Acid

The most promising adjuvant identified in the first experiment was the organosilicone-blend superspreader (Du-Wett ®). The second experiment examined the effect of a wider range of application rates of this adjuvant on the uptake of phosphorous acid over a period of 24 to 72 h.

Phosphorous acid containing [^{32}P] was applied at the equivalent rate 12 kg ha^{-1} in 100 L water in combination with various rates of the organosilicone (Table 2). Treatments were applied to the one-year-old needles of the same three clones (A, B and C) of potted *Pinus radiata* used in Experiment 1 and two treatments from that experiment were repeated. Treatment application and determination of phosphorous acid uptake were as described for Experiment 1.

Table 2: [32P] phosphorous acid uptake into*Pinus radiata*using various rates of the organosilicone-blend superspreader adjuvant (Du-Wett) at two harvest times

Treat ment no	Adjuvant concentration (%)	Harvest time HAT	[^{32}P] uptake (%)	**[^{32}P] uptake dose* (kg ha^{-1})**
2R#	none	24	3.5 d	0.42
8R#	0.20	24	43.6 abc	5.23
9	0.05	24	31.8 c	3.82
10	0.10	24	33.6 bc	4.03
11	0.10	72	47.3 ab	5.68
12	0.20	72	51.6 a	6.19

Phosphorous acid was applied at the equivalent of 12 kg ha^{-1} in all treatments. Means sharing common postscripts are not significantly different.

#The postscript R indicates a repeat treatment.

*Nominal dose calculated as (phosphorous acid rate x %_uptake).

Rolando *et al.*

Rolando *et al. New Zealand Journal of Forestry Science* 2014 **44**:8 doi:10.1186/s40490-014-0008-5

Experiment 3: Testing the Efficacy of the Best Acid/Adjuvant Combination Against *Phytophthora* Species Using a Detached Needle Bioassay

Both rates of phosphorous acid tested previously (3 kg ha^{-1} and 12 kg ha^{-1}) were evaluated with the best-performing adjuvant Du Wett ® added at 0.2% on the same three *Pinus radiata* clones (A, B and C). Treatments were applied in the equivalent of 100 L ha^{-1} using a calibrated, moving head tracksprayer (Plant Protection Chemistry NZ, Rotorua) fitted with twinjet nozzles (TJ60-8002EVS). There were a total of 27 potted plants in the trial; three for each of the two treatments (plus a control) with each clone. After treatment, the potted plants were sheltered from rain for 72 h after which they were placed outside and watered regularly.

Zoospore inoculum was prepared by growing isolates on carrot agar at 17°C for three days. Plugs of agar and mycelium were taken from the leading edge of the colonies, flooded with clarified carrot broth (Erwin & Ribeiro [1996]) and incubated for three days at 17°C. The resulting mycelial mats were rinsed thoroughly three times with sterile deionised water, drained and flooded with sterile pond water. Pond water was collected from a local pond and used in the development of*Phytophthora pluvialis* and *P. kernoviae* cultures. These were incubated in the dark for a further three days before zoospore release was induced with 45 minute intervals at 4°C in the dark at room temperature (21-22°C) on a light box. Zoospore concentrations were determined using a haemocytometer and standardised to 1×10^5 zoospores per mL. Zoospore suspensions were used within two hours of preparation.

Five days after treatment, 30 fascicles were collected from each potted plant, including the controls. Ten fascicles each, from each potted plant. were exposed to either a suspension in pond water of 1×10^5*P. pluvialis* or *P. kernoviae* zoospores per ml for 24 h. Controls were exposed to autoclaved pond water only. Following exposure, needles were placed on trays moistened with wet paper towels and incubated in a controlled environment (17°C, 65-70% relative humidity, 14 h photoperiod) for 10 days. After this time, the needles within each fascicle were separated and lesion length was measured.

Statistical Analyses

Treatments in Experiments 1 and 2 were compared by analysis of variance and least significant difference (LSD) using the statistical program Statistix 9 (Analytical Software, USA). Results from Experiment 3 (the detached needle bioassay) were analysed using the statistical programme R version 2.15.2 (R Development Core Team, 2012). A linear mixed effects model using maximum restricted likelihood estimation (R-package *nlme*) was applied to test for differences in the length of needle lesions between a zero control and phosphorous acid treatments. The fixed terms of the model contained 'phosphorous acid treatment' (factor with 3 levels: 0, 3 and 12 kg ha^{-1}) and '*Phytophthora* pathogen' (factor with 3 levels: control, *P. pluvialis* and *P. kernoviae*) and their interaction. 'Host tree identity', 'subsample identity' within tree and 'fascicle identity' within subsample were modelled as nested random effects. Tukey contrasts (R-package *multcomp*) were used to compare the effects of the phosphorous acid treatment for each *Phytophthora* species separately.

RESULTS AND DISCUSSION

Experiment 1: Effect of Four Adjuvants on Uptake of Phosphorous Acid

Comparison of Adjuvants

The results of the preliminary experiment indicated that uptake of phosphorous acid into *Pinus radiata* needles was low and proceeded very slowly. A similar uptake of phosphorous acid was recorded at 24 h after treatment for the 3 kg ha^{-1} and 12 kg ha^{-1} phosphorous acid application rates, 6.6% and 3.9% respectively (Treatments 1 & 2; Table 1). Based on this preliminary experiment, it was decided to proceed with the highest rate of phosphorous acid tested in the experiment (treatments 3–8), to maximise the acid dose absorbed by the foliage.

Addition of 0.35% pinolene as an adjuvant did not increase uptake of phosphorous acid (Treatment 3; Table 1). This was not unexpected since this type of polymer is claimed to provide increased resistance of sprays to rain wash-off and improved adhesion on difficult-to-wet plants, but is not known to increase pesticide uptake.

Addition of 0.05% alcohol ethoxylate (Treatment 4) was ineffective but the addition of the same adjuvant at 0.1% (Treatment 5) significantly increased uptake of phosphorous acid. There was no additional benefit from using a higher concentration of 0.2% (Treatment 6; Table 1). This class of adjuvant is known to improve the adhesion and spreading properties of sprays, and has been shown to promote increased uptake of a range of compounds on a variety of species (Stock et al.[1992]). These authors also showed that the magnitude of increased penetration varied according to properties of both the compound and the adjuvant and was influenced by surfactant concentration and target plant species.

The two organosilicone adjuvants (Du-Wett ® and Du-Wett ® Stainless) were used at the same rate (0.2%) (Treatments 7 & 8) and both significantly increased uptake relative to the other adjuvants tested. The organosilicone blend (superspreader adjuvant (Du-Wett ®); Treatment 8) was the most effective of all the treatments tested, increasing uptake to nine fold greater than that of the phosphorous acid used alone (Table 1). This organosilicone blend was used at a higher rate of addition than when typically used as a superspreader in horticultural operations (≤ 0.1%) in order to maximise its penetration properties.

The pH-stable organosilicone penetrant (Du Wett ® Stainless) was included in the study because it is recommended for use with acidic active ingredients (AgriMedia Ltd [2013]; Policello and Gaskin[2010]) such as phosphorous acid. However, the formulated phosphorous acid used in this study, Agrifos 600, is buffered to moderately neutral (pH of 6.8) meaning use of a pH-stable organosilicone penetrant is not essential. This was useful since the pH-stable organosilicone penetrant was more expensive than the organosilicone blend.

Standard organosilicone superspreader adjuvants should only be used at neutral pH (6–8) as these adjuvants degrade rapidly in acid/alkaline environments (Policello et al. [1995]). The superspreader adjuvant used in the current trial is disclosed as a proprietary blend of organosilicone and organic fluids and is not typically used as a penetrant. However, it contains polymers with uptake-enhancing capability when combined, at appropriate concentrations, with phosphorous acid.

Droplet Spread

The spread of droplets on a leaf surface is a measure of how well a spray can be expected to cover the plant surface following droplet adhesion. *Pinus radiata* foliage is classified as easy-to-wet, with a surface contact angle of 64° (Gaskin et al. [2005]). Surfactants are most useful for improving droplet spread and adhesion on difficult-to-wet species (Gaskin et al. [2005]). The effect of adjuvants on droplet spread on easy-to-wet foliage confirmed this (Table 1). Although significant, pinolene increased the spread of phosphorous acid spray droplets negligibly. Alcohol ethoxylate improved droplet spread and the effect increased with increasing concentration (Table 1). The pH-stable organosilicone provided slightly enhanced droplet spreading relative to either pinolene or alcohol ethoxylate. The most effective treatment, however, was the organosilicone blend superspreader, which significantly increased coverage of needle surfaces by almost three-fold compared to the controls.

Experiment 2: Optimising Uptake of Phosphorous Acid

A repeat of the control phosphorous acid treatment at a rate of 12 kg ha^{-1} confirmed that very little penetration of phosphorous acid occurred in the absence of adjuvant (Treatment 2R; Table 2). A repeat of Treatment 8 from Experiment 1 (8R, Table 2) confirmed that the organosilicone blend increased phosphorous acid uptake. This treatment and all the others using the organosilicone blend superspreader (Treatments 9–12) increased uptake by more than nine-fold relative to the control (Table 2). The highest uptake occurred at the highest concentration tested (0.2%) at both 24 h and 72 h after treatment.

Uptake with the lowest concentration of organosilicone blend (0.05%) showed high variability within replicates and, as a result, this treatment was repeated with similar variability in results. The reasons for this are not obvious, but suggest that a low concentration of organosilicone blend adjuvant will provide highly variable uptake of phosphorous acid into *Pinus radiata* foliage.

An analysis of treatments 8R, 10, 11 & 12 (Table 2) as a factorial of adjuvant concentration (0.1% and 0.2%) and harvest time (24 h and 72 h) showed that the main effects of both concentration and uptake over time were significant ($p<0.05$). Thus, uptake of phosphorous acid is expected to be greater with the higher concentration of organosilicone blend and to proceed for up to 72 HAT at least (Table 2). Droplets of all treatments were still visibly hydrated on foliage at 24 HAT, whereas at 72 HAT all droplets had visibly dried. Uptake is likely to cease when fungicide deposits become dry (Merritt [1982]) so the rate of uptake will slow with time.

Testing the Efficacy of Best Acid/Adjuvant Combination Againstphytophthora Species

Lesion length of *Pinus radiata* needles inoculated with either *Phytophthora kernoviae* or *P. pluvialis*was significantly reduced by application of the phosphorous acid treatments containing 0.2% organosiliconeblend adjuvant at both application rates tested compared to controls (Figure 1). However, there was no significant difference in lesion length between the low- and high-dose treatments used and either *Phytophthora* species tested. This result indicated that sufficient phosphorous acid had been absorbed by the needles to affect pathogen activity at both concentrations. In addition, these results indicate that phosphorous acid has activity against the target *Phytophthora* species when it is applied with organosilicone blend adjuvant. Similar bioassays have been used extensively to demonstrate the efficacy of phosphorous acid application to imply field based control for a range of *Phytophthora* diseases (Barrett et al. [2003]; Garbelotto and Schmidt [2009]; Shearer et al. [2012]). While bioassays can provide a useful indication of treatment efficacy, in some cases, they have also been shown to differ significantly from field based analyses and are therefore no substitute for field based results (Shearer et al. [2012]). Therefore, field trials involving application of phosphorous acid to mature *Pinus radiata* trees need to be carried out to determine an optimum dose for management of needle diseases in the field.

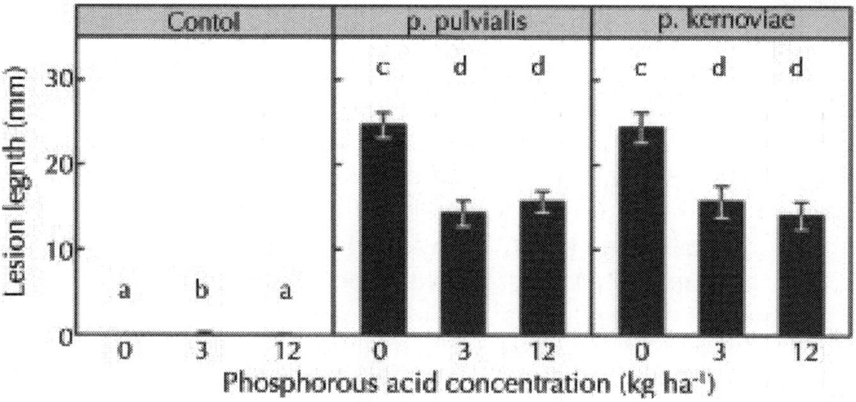

Figure 1: Lesion length (model predictions and confidence intervals) on*Pinus radiata*needles exposed to*Phytophthora pluvialis*or*Phytophthora kernoviae*for 24 hours following treatment with either 3 kg ha-1or 12 kg ha-1phosphorous acid each applied in 100 L water containing 0.2%organosilicone-blend superspreader adjuvant (Du-Wett ®).

Kanaskie et al. ([2011]) conducted field trials to test the efficacy of phosphorous acid for the control of sudden oak death in south western Oregon tanoak. In aerial spray trials phosphorous acid was applied at two rates, 17.36 kg ha^{-1} and 34.72 kg ha^{-1}, in conjunction with Pentra-bark ® in 160 L water ha^{-1}. Detached-twig and log bioassays with *P. ramorum*, as well as *in situ* stem inoculations with *Phytophthora gonapodyides* (Petersen) Buisman, were carried out to investigate treatment efficacy. Aerial spray with phosphorous acid at both application rates consistently resulted in smaller bole lesions on trees challenge inoculated with *P. gonapodyides in situ* and logs inoculated with *P. ramorum*. However, results from detached canopy twig assays were variable and showed only small treatment effects. These findings demonstrate the potential variability of field based treatments and the need to assess treatment effects both in-field and using bio-assays tested across different tissues. In the present study, bioassays focussed on detached needle inoculations as red needle cast has, as yet, only been shown to affect *Pinus radiata* needles (Dick et al. [2013]). Further confirmation of these results in the field is essential to establish the efficacy of broad-scale treatment for the control of red needle cast in *Pinus radiata*.

CONCLUSIONS

There was very little penetration of phosphorous acid into the foliage of *Pinus radiata* 24 hours after treatment in the absence of adjuvants. The addition of either an alcohol ethoxylate or organosilicone adjuvant improved uptake. The highest uptake of phosphorous acid occurred with an organosilicone blend adjuvant (Du-Wett ®) at a concentration of 0.2%, and uptake progressed over a period of more than 24 hours after treatment. The addition of either of these adjuvants improved surface coverage of phosphorous acid sprays and they are also likely to have some beneficial effects on spray retention.

It is clear that the use of an organosilicone blend adjuvant in phosphorous acid treatments will improve coverage of sprays on *Pinus radiata,* in addition to its effects on the penetration of phosphorous acid into the plants. Its use is recommended to increase the uptake of phosphorous acid when applied to *Pinus radiata* for management of red needle-cast disease. Preliminary indications from the detached needle bioassays indicate that the rates of phosphorous acid used in this investigation (3 kg ha^{-1} and 12 kg ha^{-1}) have the potential to reduce infection of *Pinus radiata*by *Phytophthora* species including *P. kernoviae* and *P. pluvialis*. Field trials are needed to confirm this.

AUTHORS' CONTRIBUTIONS

CR and RG lead the conceptual development of this paper. RG and DH carried out all experimental work to characterise the penetration of phosphite as affected by four commercially available adjuvants. CR designed the study to test the best performing treatment on potted *P. radiata,* with NW the conceptual lead on work associated with in-vitro testing of *P kernoviae* and *P. pluvilais*. MB analysed data and developed models to present Figure 1. All authors read and approved the final manuscript.

ACKNOWLEDGEMENTS

The Forest Owners Association, New Zealand, and Ministry of Business Innovation and Employment, New Zealand, provided the funding for this work.

REFERENCES

1. (2013). New Zealand Novachem Agrichemical Manual. Ashcroft House, Christchurch, New Zealand.
2. Barrett, S, Shearer, B, Hardy, GESJ (2003). The efficacy of phosphite applied after inoculation on the colonisation of *Banksia brownii* sterns by *Phytophthora cinnamomi. Australasian Plant Pathology, 32*, 1–7.
3. Cohen, Y, & Coffey, MD (1986). Systemic fungicides and the control of oomycetes. *Annual Review Phytopathology, 24*, 311–388.
4. Dick, M, Williams, NM, Hood, IA, Bader, M. (2014). Pathogenicity of Phytophthora pluvialis to Pinus radiata and its relation with red needle cast disease in New Zealand. New Zealand Journal of Forestry Science.44, 6.
5. Erwin, DC, & Ribeiro, O (1996). Phytophthora Diseases Worldwide. APS Press, Minnesota.
6. Garbelotto, M, & Schmidt, DJ (2009). Phosphonate controls sudden oak death pathogen for up to two years. *California Agriculture, 63*, 10–17.
7. Gaskin, RE, Steele, KD, Forster, WD (2005). Characterising plant surfaces for spray adhesion and retention. *New Zealand Plant Protection, 58*, 197–183.
8. Gaskin, RE, Logan, DP, May, W, Rowe, CA, Steele, KD, van Leeuwen, RM, Connolly, PG (2012). Control of passionvine hopper and cicada eggs on kiwifruit canes with bifenthrin and a new super-penetrant adjuvant. *New Zealand Plant Protection, 65*, 100–105.

9. Graham, JH (2011). Phosphite for control of Phytophthora diseases in citrus: model for management of *Phytophthora* species on forest trees? *New Zealand Journal Forestry Science, 41S*, S49–S56.
10. Hardy, GESJ, Barrett, S, Shearer, BL (2001). The future of phosphite as a fungicide to control the soilborne plant pathogen *Phytophthora cinnamomi* in natural ecosystems. *Australasian Plant Pathology, 30*, 133–139.
11. Kanaskie, A, Hansen, EM, Sutton, W, Reeser, P, Choquette, C (2011). Application of phosphonate to prevent sudden oak death in south-western Oregon tanoak (*Notholithocarpus densiflorus*) forests. *New Zealand Journal of Forestry Science, 41S*, S177–S187.
12. Lobato, MC, Machinandiarena, MF, Tambascio, C, Dosio, GAA, Caldiz, DO, Daleo, GR, Andreu, AB, Olivieri, FP (2011). Effect of foliar applications of phosphite on post-harvest potato tubers. *European Journal of Plant Pathology, 130*, 155–163.
13. Merritt, CR (1982). The influence of form of deposit on the phytotoxicty of MCPA, paraquat and glyphosate applied as individual drops. *Annals of Applied Biology, 101*, 527–532.
14. Policello, GA, & Gaskin, RE In Baur P, Bonnet M (Eds.) (2010). Hydrolysis resistant organosilicon adjuvant for glyphosate. *Proceedings of the 9th International Symposium on Adjuvants for Agrochemicals Techinical University of Munich* Freising, Germany.
15. Policello, GA, Stevens, PJG, Forster, WA, Murphy, GJ In Hall FR, Berger PD, Collins HM (Eds.) (1995). The influence of pH on the performance of organosilicone surfactants. *Pesticide Formulations and Application Systems: 14th Volume* (pp. 313–317). ASTM, West Conshohocken, PA, USA.
16. Rebollar-Alviter, A, Madden, LV, Ellis, MA (2007). Pre- and post-infection activity of azoxystrobin, pyraclostrobin, mefenoxam, and phosphite against leather rot of strawberry, caused by*Phytophthora cactorum*. *Plant Disease, 91*, 559–564.
17. Reeser, PW, Sutton, W, Hansen, EM. (2013). Phytophthora pluvialis, a new species found in mixed tanoak – Douglas-fir forests of western Oregon, U.S.A. North American Fungi 8(7), 1–8.

18. Reglinski, T, Spiers, TM, Dick, MA, Taylor, JT, Gardner, J (2009). Management of phytophthora root rot in radiata pine seedlings. *Plant Pathology, 58*, 723–730.
19. Shearer, BL, & Crane, CE (2009). Influence of site and rate of low-volume aerial phosphite spray on lesion development of *Phytophthora cinnamomi* and phosphite persistence in *Lambertia inermis*var. *inermis* and *Banksia grandis*. *Australasian Plant Pathology, 38*, 288–304.
20. Shearer, BL, Crane, CE, Scott, PM, Hardy, GES (2012). Variation between plant species of in-planta concentration and effectiveness of low-volume phosphite spray on *Phytophthora cinnamomi* lesion development. *Australasian Plant Pathology, 41*, 505–517.
21. Stock, D, Edgerton, BM, Gaskin, RE, Holloway, PJ (1992). Surfactant-enhanced foliar uptake of some organic compounds: Interactions with two model polyoxyethylene aliphatic alcohols. *Pesticide Science, 34*, 233–242.
22. Tomlin, C (2006). The Pesticide Manual. British Crop Protection Council, Hampshire.
23. Tynan, KM, Wilkinson, CJ, Holmes, JM, Dell, B, Colquhoun, IJ, McComb, JA, Hardy, GEJ (2001). The long-term ability of phosphite to control *Phytophthora cinnamomi* in two native plant communities of Western Australia. *Australian Journal of Botany, 49*, 761–770.
24. Zabkiewicz, JA (2000). Adjuvants and herbicidal efficacy - Present status and future prospects.*Weed Research, 40*, 139–149.

Citations

CHAPTER 1

Janna L. Beckerman (2013). Detection of Fungicide Resistance, Fungicides - Showcases of Integrated Plant Disease Management from Around the World, Dr. Mizuho Nita (Ed.), ISBN: 978-953-51-1130-6, InTech, DOI: 10.5772/55981.

CHAPTER 2

Marina D. Soković, Jasmina M. Glamočlija, and Ana D. Ćirić (2013). Natural Products from Plants and Fungi as Fungicides, Fungicides - Showcases of Integrated Plant Disease Management from Around the World, Dr. Mizuho Nita (Ed.), ISBN: 978-953-51-1130-6, InTech, DOI: 10.5772/50277.

CHAPTER 3

Maria Celeste Dias, "Phytotoxicity: An Overview of the Physiological Responses of Plants Exposed to Fungicides," Journal of Botany, vol. 2012, Article ID 135479, 4 pages, 2012. doi:10.1155/2012/135479.

CHAPTER 4

Carla D. Garzon and Francisco J. Flores (2013). Hormesis: Biphasic Dose-Responses to Fungicides in Plant Pathogens and Their Potential Threat to Agriculture, Fungicides - Showcases of Integrated Plant Disease Management from Around the World, Dr. Mizuho Nita (Ed.), ISBN: 978-953-51-1130-6, InTech, DOI: 10.5772/55359.

CHAPTER 5

Chao Yang, Chantal Hamel, Vladimir Vujanovic, and Yantai Gan, "Fungicide: Modes of Action and Possible Impact on Nontarget Microorganisms," ISRN Ecology, vol. 2011, Article ID 130289, 8 pages, 2011. doi:10.5402/2011/130289.

CHAPTER 6

M. H. Rashid, M. Ashraf Hossain, M. A. Kashem, Shiv Kumar, M. Y. Rafii, and M. A. Latif, Efficacy of Combined Formulations of Fungicides with Different Modes of Action in Controlling Botrytis Gray Mold Disease in Chickpea, doi: 10.1155/2014/639246.

CHAPTER 7

Roozbeh Hushiarian, Nor Azah Yusof, and Sabo Wada Dutse, Detection and control of Ganoderma boninense: strategies and perspectives, doi:10.1186/2193-1801-2-555.

CHAPTER 8

Michael Faust, Carolina Vogs, Stefanie Rotter, Janina Wöltjen, Andreas Höllrigl-Rosta, Thomas Backhaus, and Rolf Altenburger, Comparative assessment of plant protection products: how many cases will regulatory authorities have to answer?,.doi:10.1186/s12302-014-0011-8.

CHAPTER 9

Carol Rolando, Robyn Gaskin, David Horgan, Nari Williams, and Martin K-F Bader, The use of adjuvants to improve uptake of phosphorous acid applied to Pinus radiata needles for control of foliar Phytophthora diseases, doi:10.1186/s40490-014-0008-5.

Index